END IN FIRE

The supernova in the Large Magellanic Cloud

END IN FIRE

The supernova in the Large Magellanic Cloud

PAUL MURDIN
Royal Greenwich Observatory

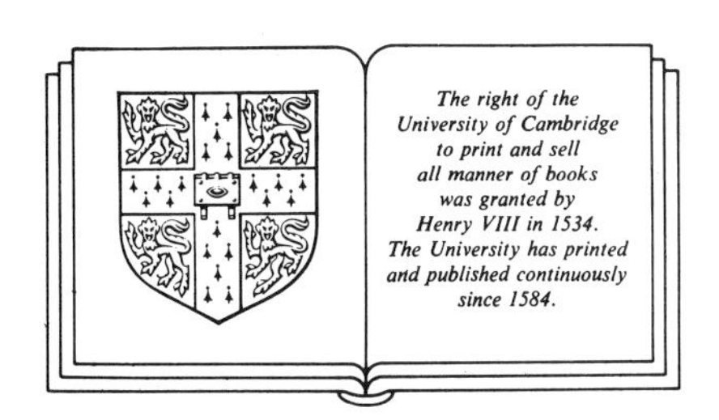

CAMBRIDGE UNIVERSITY PRESS
Cambridge
New York Port Chester
Melbourne Sydney

Published by the Press Syndicate of the University of Cambridge
The Pitt Building, Trumpington Street, Cambridge CB2 1RP
40 West 20th Street, New York NY10011, USA
10 Stamford Road, Oakleigh, Melbourne 3166, Australia

First published 1990

Printed and bound in Great Britain by
Billing & Sons Ltd, Worcester

British Library Cataloguing in Publication Data
Murdin, Paul
End in fire.
1. Supernovae: Supernova 1987A
I. Title
523.8′446

Library of Congress Cataloguing in Publication Data

Murdin, Paul
End in fire, the supernova in the large magellanic cloud
Paul Murdin.
p. cm.
Bibliography p.
ISBN 0-521-37495-2
1. Supernovae. 2. Magellanic Clouds.
QB843.S95M86 1989
523.8′446—dc20 89-3577 CIP

ISBN 0 521 37495 2

Contents

Preface

This book is about the nearest, brightest supernova for 400 years. Although supernovae are rare, their properties are so dramatic that they can be visualised quite readily from a broad brush picture. The science which goes into interpreting the observations of supernovae and creating theories about them is exciting, very varied, complicated in detail but surprisingly simple in outline. So the book is about the broad picture which science shows of the supernova.

I have also enjoyed telling about the astronomers who studied the supernova. Not everything in their journeys of discovery went smoothly and straightforwardly. A journey of discovery is an adventure: if you knew how it was going to turn out, it would not be exciting. Astronomers make mistakes, get jealous, get lucky, jump to conclusions, compete, boast, work hard, succeed, fail, laugh and cry. I never met a good astronomer who fitted the ivory-tower, white-coat stereotype. So the book is also about what astronomers do.

The book deals at once with very large and very small numbers; this is because it deals with a subject which represents at the same time supernovae, the most powerful explosions in the universe, and neutrinos, the least massive and least interactive fundamental particles. Part of the fun is relating these numbers and deriving some of the properties of supernovae from them.

I have enjoyed giving numerical estimates of various things. Astronomers call these 'back of an envelope' calculations, because they are usually scribbled on whatever is handy, including envelopes, but not excluding paper napkins, the margins of typescripts, old copies of the *Astrophysical Journal* and pompous notices ripped from the noticeboard. To me it is marvellous that these scribbles get anywhere near a representation of the truth, just as it is astonishing that a simple and rapid line drawing by a Picasso, can, in a sketch, convey someone's character.

If astronomy is defined as what astronomers do, say and think, then this definition allows me to be inconsistent with physical units, because astronomers are. I was brought up on British units, educated in cgs units and have since learnt some of the mks (SI) system. Professionally, I also have to handle units in none of these systems, such as parsecs, angstroms and solar masses. I am not particularly different from most astronomers in that we have to slip relatively easily from one set of units to another, otherwise we cannot comprehend the relative dimensions of things.

In the same way I have allowed myself to be intuitive in the mathematical calculations. Astronomers when they chat are like that. Backs of envelopes are too small for theorems.

I write dates and times in the astronomical convention, e.g. 1987 February 23 07:35 UT: the most significant digits come first, just as if the date was a number.

If this book was a text book I would have to try to be rigorous. It is not, and I have not been. But I hope the book is not too sloppy.

If this book was a comprehensive scientific review I would have carefully to give all sides of an argument, referring to all the papers. It is not that either, and so I have not done so. But I hope the book is not too impressionistic.

I want in this book to make the theory behind the supernova of 1987 accessible to the largest possible audience. I have assumed that the reader has a school education in physics. If you find something that is too hard, skip on to the next section.

Parts of this book are based upon articles which I have written for *Contemporary Physics, The Physics Bulletin, Nature, New Scientist*, the *Quarterly Journal of the Royal Astronomical Society* and *Astronomy Now*. Parts are also based on editions of BBC TV's *Sky at Night*, presented by Patrick Moore and produced by Pieter Morpurgo, and *Tomorrow's World*, produced by Dana Purvis, on ITN's *Channel 4 News* and *News at Ten*, presented by Lawrence McGinty, on BBC Radio 4's *Science Now* presented by Peter Evans and Georgina Ferry, and on programmes of the BBC World Service's Science Unit, presented by Peter Beer and Martin Redfearn. I have developed the material in talks given at universities, laboratories and astronomy societies throughout Britain.

I am grateful to the editors of the journals, the producers and presenters of the programmes and the organisers of the lectures for the opportunities which they gave me to develop this book. I am very grateful for the way in which the professionals with whom I worked taught me to express the story. I am also grateful to the many people in the audiences for my talks who challenged me with perceptive questions and provocative comments, and who shared with me their enthusiasm for this knowledge and their interest in the science of astronomy which I love.

I am grateful to several friends and colleagues for reading parts or all of the manuscript and supplying illustrations: W.D. Arnett, E. Budding, A. Cassatella, R. Chevalier, C. Henshaw, J. Jelley, M. Karovska, M. Mackenzie, D. Malin, R.N. Manchester, R. McNaught, P. Meikle, S. Metz, H. Ogelman, R. Olowin, S. Sakhar, M. Turner, W. Wamsteker, P. Willmore, A. Wolfendale.

This book was written on my Atari ST word processor. Parts were written in the control room of the William Herschel Telescope on La Palma while taking long integrations or waiting for the weather, and I am grateful to Andrew Pickles for allowing me to use his ST for the purpose, even when it meant that he, poor chap, had to use the VAX 8300 instead.

Herstmonceux Paul Murdin
September 1988

1
Introduction

A nova (literally 'new star') is an event which represents the sudden brightening of a star (a bright star appears where no star was particularly noticeable before). It is thus an energetic outburst. Supernovae are similar but extremely energetic outbursts. Indeed, in terms of the power which they radiate, they are the most energetic explosions known, apart from the unique occasion at the start of the Universe, known colloquially as the Big Bang.

In a supernova an energy of about 10^{46} J is liberated in a few seconds. The total power in all forms, released in those seconds, is equal to the luminosity of the rest of the Universe. Part of the energy is in the form of radiation – light and infrared radiation – which is released over a timescale of months, and the radiated energy of 10^{36} J/s may cause a supernova to equal the brightness of all the stars in its parent galaxy put together. A supernova rises to its maximum brightness very quickly, typically on a timescale of a day, and fades more slowly, typically on a timescale of months.

In an average galaxy like our own, a supernova occurs at random every 50 years or so. In spite of the enormous output of energy, most supernovae which explode in our Galaxy seem to remain undiscovered, hidden in dust or confused amongst brighter-seeming nearby stars. The vast majority of supernovae have been discovered telescopically in galaxies outside our own.

Up to February 1987, 620 supernovae had been recorded. The last supernova seen in our Galaxy, and the last which was seen by an unaided human eye, occurred in 1604; it was studied by Johannes Kepler and goes under his name. The previous supernova visible by the unaided eye was in 1572 and is named for Tycho Brahe. It is considered by astronomers just pure bad luck that there were two bright supernovae in 30 years in the Renaissance but not another for the 383 years since then.

In spite of their rarity, supernovae shape human destiny. Supernovae create the elements of which our own material world is moulded. They scatter these elements as seeds for the construction of new solar systems, including new planets, new life forms. Material in our own selves was once inside supernovae of the distant and unknowable past. If we study supernovae, we study our own human origins.

Or perhaps astronomers are the way which supernovae have of learning about themselves.

Supernovae contribute to the cosmic rays which help drive the evolution of genetic material. Supernovae also drive the evolution of systems as large as galaxies.

On 1987 February 23, there occurred the brightest supernova for nearly 400 years. It was easily visible to the unaided eye, but only to people who live in the Earth's southern hemisphere.

Fig. 1. David Malin combined a positive (white stars) image of a UK Schmidt Telescope plate showing the supernova with a negative of an earlier plate showing the progenitor (black star at the centre of the saturated supernova image). The 30 Doradus Nebula shows grey, and in a bas relief, due to the residual misalignment of the two plates. (© 1987 Royal Observatory, Edinburgh.)

The supernova is known in the dry, catalogue jargon of scientists as SN 1987A. It occurred in the Large Magellanic Cloud (LMC), a neighbour galaxy to our own. Distant about 170 000 light years from the Earth, the LMC is one of two objects, news of the existence of which was brought to Europe after the round-the-world voyages of the explorer Magellan. They were thought at that time to be pieces of the Milky Way which had detached themselves. Its smaller sister, the Small Magellanic Cloud (SMC), is slightly further away from us. Together the Magellanic Clouds are satellites of our Galaxy (an analogy would be our Earth having two moons).

In cosmic terms, the Magellanic Clouds are on our doorstep. They are five times closer than the Andromeda Galaxy, the nearest galaxy to us of the same size as our own Galaxy. They are much smaller than the Andromeda Galaxy: the LMC has 30 times fewer stars, the SMC over 100 times fewer.

The LMC supernova is the brightest supernova for nearly 400 years, and probably the nearest for that long too. It is the first supernova in which the progenitor star (the seat of the explosion) has been properly identified. It appeared to us as the brightest star ever noted in an external galaxy. It is the first nearby supernova studied with scientific equipment (Kepler's supernova occurred a few years before the invention of the telescope!). It is the first celestial object from which neutrinos have been identified, apart from our Sun. It is the first in which we have seen elements being created, and the radioactivity of these newly-formed elements released the first gamma-ray spectral lines seen from a celestial object outside the Sun.

The LMC supernova is the best and most critical test of the calculations made by supernova theoreticians. In spite of some surprises they have performed very well in the face of the challenge of the new observations. The confirmation of their theories is a triumph of the human imagination, in which we can all take pride. Fellow human beings in our own civilisation and history have been able accurately to put together slender clues from physics and astronomy in a consistent and accurate theory.

With his tongue in his cheek, Kepler said of the supernova of 1604 that it would being good fortune to publishers, at least, since it would bring a spate of pamphlets and books. The modern equivalents of the pamphleteers are the scientists who have rushed into print with their ideas about the LMC supernova. By the end of April 1988, when the supernova was just over a year old, the list of publications devoted to it numbered at least 375.

The supernova received a recognition which is reckoned to be a distinction of sorts by appearing on the cover of *Time* magazine (*Time* 1987). The supernova was clearly the astronomical event of the year, some say of the decade, or the century.

It has been possible to study the supernova in the LMC in more detail than any other supernova, and to get to grips with processes which shaped our own destiny. The supernova has thrown light into areas of astronomy and even of physics where there were at least shadows, if not darkness. The LMC supernova will have become, a generation from now, the supernova which made the subject classical.

The typescript of this book has been completed just over a year and a half after the supernova was announced, and last-minute additions were made in mid-1989. I have tried to put together a consistent and understandable story about it, and to share my excitement in it. The explosion is continuing and no doubt there will be surprises in store for us over the coming years.

2
The first two days: discovery – and prediscovery

Supernovae occur suddenly and without warning. They are discovered only by chance.

Shelton's discovery of the supernova

The supernova was discovered on 1987 February 24 by Ian Shelton. Shelton is a University of Toronto astronomer, a research assistant who was assigned to Las Campanas Observatory. At 2500 m, it sits high on one of the Andes mountains of the Atacama Desert in northern Chile. It is operated by the Mt Wilson Observatory and is the southern observing station of the Carnegie Institute of Washington DC (Plate 2).

Shelton had taken a three hour photographic exposure of the LMC with the Bruce 10 in telescope, using photographic materials cadged from another project. The photograph was part of what he envisaged as his new routine observing programme to look for variable stars and novae in the Magellanic Clouds. He had begun the programme only three nights beforehand, after badgering the resident scientist at Las Campanas Observatory to let him try the programme out on the idle 10 in telescope.

Before going to bed, Shelton developed the photographic plate and, at 05:40 UT†, lifting the photograph from the developing tank, he saw an unfamiliar spot. At first he was sure that it was a plate flaw, giving the impression of a star. When Shelton examined the spot carefully, however, he

† UT stands for Universal Time, the time system adopted by astronomers to avoid any confusion of time zones in telling a 'universal' story, like the one following. UT is identical to Greenwich Mean Time.

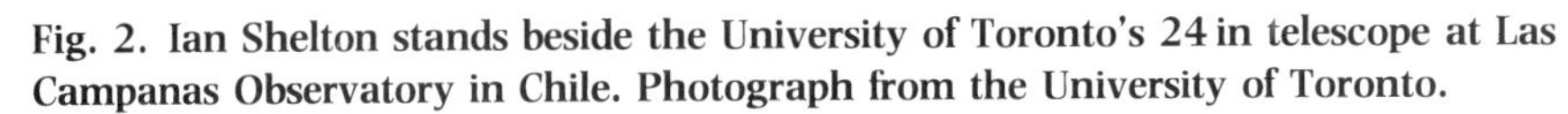

Fig. 2. Ian Shelton stands beside the University of Toronto's 24 in telescope at Las Campanas Observatory in Chile. Photograph from the University of Toronto.

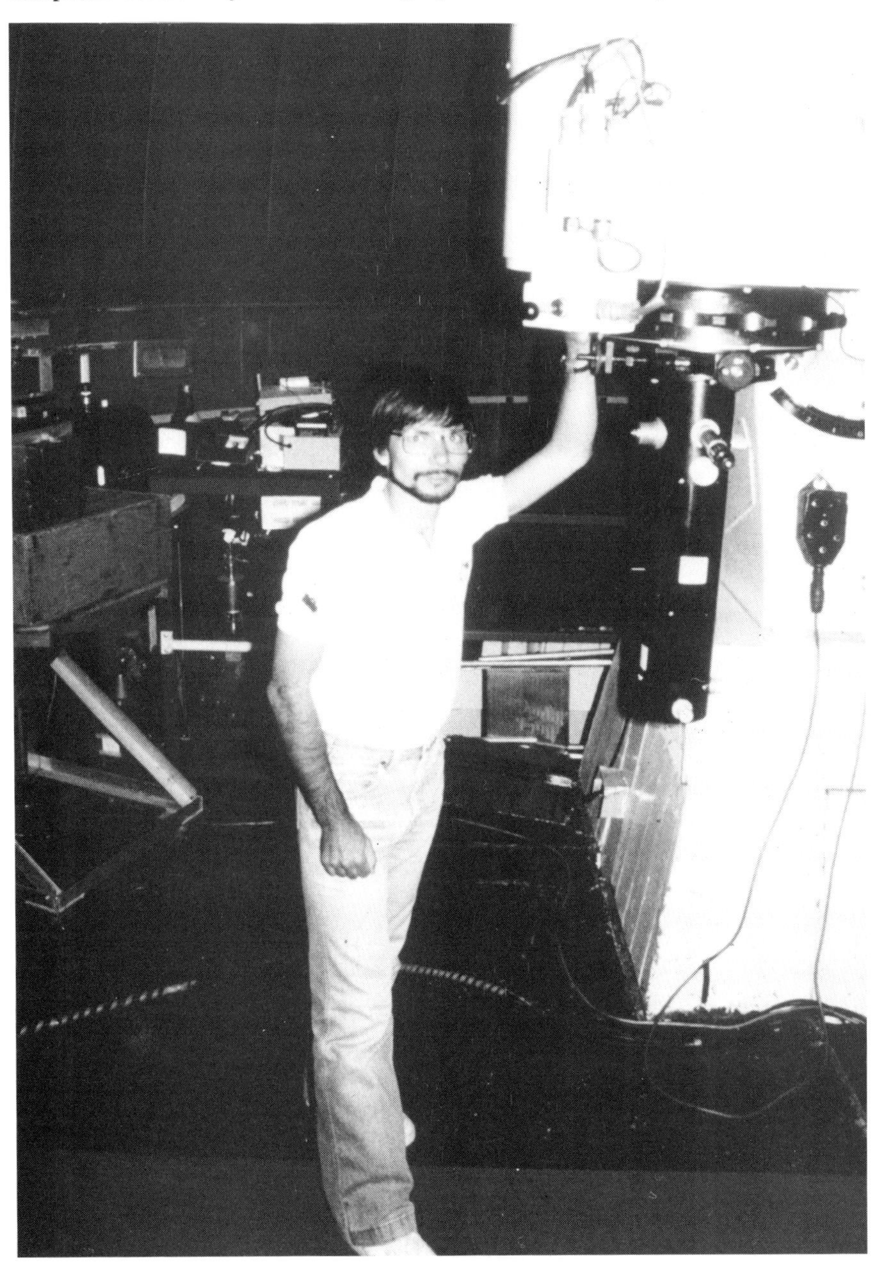

found that it was no flaw: it looked like a fifth magnitude star. Shelton wondered why he had not used it as a guide star. While taking photographs, astronomers view a guide star through an auxiliary eyepiece. This is to correct irregularities in the motion of the main telescope while it is exposing the photograph on the sky.

A star this bright would have been a convenient one with which to guide the telescope. Then Shelton realised the fifth magnitude star should not be there. He walked outside, looked up at the LMC and saw that the star was really there and must be a nova – or a supernova!

Shelton was the first person knowingly to record the supernova; he did not yet know that in observatory records would be found evidence of the explosion a day before he discovered it.

Further discoveries

As recorded by Jedrzejewski (1987), Shelton's reaction was to go to the dome of the 40 in telescope of Las Campanas Observatory and share his discovery, and the thoughts in his mind. The two visiting astronomers in the windowless control room of the 40 in were Robert Jedrzejewski and Barry Madore, and their Chilean night assistant was Oscar Duhalde (pronounced 'Dwolder'), whose job it was to operate the telescope for them.

Shelton asked about the magnitudes of novae in the LMC and what it would mean if he had discovered a star which on the previous night had been magnitude 12, but was now magnitude 5. Madore said that it must be a supernova.

'At this point,' wrote Jedrzejewski, 'Oscar chipped in and confirmed that he had sighted the object visually earlier in the evening.'

Duhalde had noticed it as he strolled about the mountain top, enjoying the splendid view of the southern stars. He had taken a break from his work, leaving the telescope to track automatically; he had brewed a cup of coffee and stepped outside to glance at the sky at about 03 : 00 UT. He noticed that the 30 Doradus Nebula in the LMC was much brighter than usual. As he turned away, he said to himself '30 Doradus *that* bright?' and looked back carefully: 'I saw the real 30 Doradus and I knew this object wasn't there before.' When he returned to work, an impatient astronomer spoke before he could talk about the new star, asking Duhalde to move the telescope to a new object, and Duhalde forgot to mention the phenomenon (Maran 1988).

When they realised the significance of the new star, the whole group went outside the dome to see the supernova. To their dismay, they had realised too late that there was a supernova: the 100 in telescope could not point to the supernova – it was too low in the sky. Moreover, the 40 in telescope on which they were working had too sensitive an instrument to record such a bright star (see Chapter 3). Duhalde made a simple visual estimate of the supernova's brightness (Table 1).

As night-time moved westwards across the Pacific Ocean, a renowned New Zealand amateur spotted the supernova. The sky was clearing as Albert Jones of Nelson, NZ, left a committee meeting for the more interesting job of observing variable stars in the LMC. He observed the region near to the Tarantula Nebula and his attention was taken by a bright star near to the central part of the field of his 30 × 78 finder telescope at about 08:53 UT. His knowledge of the southern sky was so good that he recognised immediately that the star was new. He estimated its magnitude at 5.6–7.0, but through cloud it must have been difficult to be sure.

Fig. 3. Shelton took two pictures of the LMC with the 10 in refractor of Las Campanas Observatory on the day before and the day after the supernova explosion. In the first picture, the white arrow marks the progenitor star Sk – 69 202, four hours before the collapse of its core. It looks the same as it has looked for the duration of human history, one of many stars of comparable faintness in the LMC whose 'bar' cuts diagonally across the picture. In the second photograph, taken 21 hours after core collapse, Sk – 69 202 has exploded to more

(*a*) 1987 February 23 03 : 36 UT

Jones telephoned Frank Bateson, who organises observations of variable stars by amateur astronomers in New Zealand. He alerted Brian Moreno and Stan Walker, two amateurs at the observatory in Auckland, NZ, who measured it with a photoelectric photometer near to 11 : 00 UT. Bateson also passed the message to the Anglo–Australian Telescope (AAT) at Coonabarabran, NSW. The night assistant that night, Kevin Cooper, alerted other observers on Siding Spring Mountain, including Tom Cragg, a well-known amateur astronomer and another AAT night assistant, and Rob McNaught of the Royal Greenwich Observatory (RGO). McNaught operates the southern hemisphere satellite tracking camera for the RGO. He is also well known for his longstanding interest in variable star observations, even as an amateur astronomer, before he left England. McNaught was already clamping his camera ready to take his nightly photo of the LMC through light cirrus as the phone rang. He started the exposure and went to answer the phone. Hearing the news, McNaught looked outside and confirmed the supernova in seconds!

than the size of Saturn's orbit around the Sun and brightened several hundred-fold. It has become the brightest star on this photograph, and accounts for 10% of the total brightness of all the stars recorded here. (The photographic process, designed to record the faint stars at the expense of the appearance of the bright ones, emphasises the patchy nebulae and star clusters above their natural importance.) © 1987 University of Toronto.

(*b*) 1987 February 24 04 : 48 UT

Fig. 4. Two pictures taken with the UK Schmidt Telescope in Australia show the 30 Doradus Nebula, known as 'the Tarantula' because of its spidery shape. The lower photograph was taken on 1987 February 27 at 10:13 UT and SN 1987A shines brightly with a halo and cross superimposed on its image. (These are happy, accidental consequences of artefacts in the telescope.) The upper photograph was taken four years before the supernova. © Royal Observatory Edinburgh.

Astrogram

When they realised that Shelton and Duhalde had discovered a supernova, the astronomers at Las Campanas tried to notify the International Astronomical Union's (IAU) Central Bureau for Astronomical Telegrams. The Bureau, founded 100 years ago, has the job of acting as a clearing house for astronomical information which must be acted upon quickly. But Madore could not at first get through on the telephone from Las Campanas to the Central Bureau in Massachusetts – there was a problem on the international telephone line. The night assistant for the 100 in telescope, Angel Guerra, drove in the morning from Las Campanas to the town of La Serena to send the message by telex. Later the phone problem cleared up, and Chilean astronomers were able to talk to Brian Marsden. Because of the delay, McNaught got through to the Central Bureau while one of them was talking on the other line; he passed on information which he had gathered that night.

The supernova was designated SN 1987A. Astronomers label supernovae (of which they typically discover a dozen per year, usually in faint, distant galaxies) by the year of discovery, and in order, with the capital letters A, B, C, ... If in a given year more than 26 supernovae are discovered they start a double lettering system with lower case letters aa, ab, ac, ... This supernova was the first one discovered in 1987 – it was SN 1987A.

Daniel Green of the Bureau immediately notified astronomers worldwide of the discovery. The 'astrogram' read as follows:

SHELTON DUHALDE MCNAUGHT LMC 1987A SUPERNOVA SHELTON
19871 70224 333// 05354 16916 03045 48710 25315
BRIGHTENING 0.5 MAGNITUDE IN FIVE HOURS
GREEN 4FEB24/1500Z

First came the names of the three astronomers, Shelton, Duhalde and McNaught, who are credited in this astrogram with the discovery. McNaught was not in fact a codiscoverer; he had phoned the Central Bureau with his observations of the supernova in the belief that Bateson had already communicated Jones' discovery: Bateson had not, and McNaught did not think to mention it. Instead, the Central Bureau jumped to the conclusion that McNaught had independently found the supernova.

Then came the name of the galaxy in which the supernova appeared, the LMC. There followed the supernova designation, 1987A. Next is the name of the observer who supplied the information.

A series of numerical five-figure groups encode the data about the supernova which astronomers require:

> 19871: The position of the supernova is defined in telescope coordinates at the date of measurement (the first group of digits gives the year of the coordinates, 1987, and a '1' implying that the position is approximate).
> 70224 333//: This is the date of the observation: [198]7 02 24 is February 14, 1987, and the time is 08:00 UT (0.333 of a day – the slashes mean nothing and simply make up a group of five characters).
> 05354 16916: The position is Right Ascension 05 h 35.4 min. Declination $-69°$ 16.0′ (the leading digit '1' in the fifth group represents a minus sign).
> 03045: The last group of data is the magnitude of the supernova: it was visual magnitude 4.5.
> 48710 25315: Two final numeric groups are checksums (the last five figures of the arithmetic sum of the data to check its consistency during transmission). The first checksum checks all the numerical data $(19871+70224+33300+05354+16916+03045=148710$, and we drop the most significant digit because it makes the checksum six digits, one too many). The second checks the position and magnitude $(05354+16916+03045=25315)$.

Before the last line, which gives the sender of the astrogram and its timing on February 24 at 15:00 UT (Z in the telegram stands for Zulu, the alphabetical code for the GMT time-zone), Green noted that the visual estimates of the supernova's brightness indicated that it was still brightening, rather rapidly.

During the next few hours the identical telegram was being transmitted individually to each of the numerous observatories which subscribe to this IAU service. The bald code arrived on the telex machines, and often waited hours for someone to notice the announcement of the brightest supernova for 380 years. Observatories which could not afford, or did not need, this telegram service were informed of the supernova by the Circulars, mailed postcards which are used to back-up telegrams with more extensive comment.

In parallel with the Central Bureau's official communications, the unofficial jungle drums of astronomy began to beat: astronomers who had heard the news – from Las Campanas, from the University of Toronto, from Marsden – passed the message.

In later months the news would become to astronomers like the news of the

Kennedy assassination: they remembered what they were doing when they heard of it. A typical story is Stephen Maran's. He was walking down a hallway of the Goddard Space Flight Center (GSFC) towards the Coke machine, and overheard the excited comment: 'It's the worst one in 300 years.' He wondered what volcanic disaster had struck, but had misheard: it was not the 'worst', but the 'first'.

International Ultraviolet Explorer

The International Ultraviolet Explorer (IUE) satellite is an orbiting observatory with an 18 in telescope used to obtain ultraviolet spectra (1150–3200 Å) of celestial objects. It was launched in 1978 – in space terms it is fairly described as ageing. It is jointly operated by the US National Aeronautics and Space Administration (NASA) from GSFC near Greenbelt, Maryland, and by the European Space Agency and the UK Science and Engineering Research Council from a tracking station at Villafranca del Castillo (Vilspa) near Madrid. The US and European organisations share the use of the satellite in the ratio 2/3 to 1/3 and astronomers from North America and Europe cooperate, and compete, in its use.

The satellite has so-called 'target of opportunity' programmes. These rules allow the tracking stations to depart from the agreed plans for the satellite's use and to target observations on unexpected and significant events which provide opportunities for discoveries to be made.

The LMC supernova certainly qualified for the 'target-of-opportunity' designation.

Both the US and Europe had set up target-of-opportunity teams for supernovae – they had already achieved spectacular results on supernovae which, in distant galaxies, were many thousands of times fainter. When the LMC supernova was announced, the IUE teams were prepared with contingency plans for just such an event. IUE's telescope was re-directed towards it: the satellite was tracking the supernova by about 19:00 UT and the first ultraviolet spectrum of the supernova was obtained from the GSFC at 20:04 UT by George Sonneborn, the IUE resident astronomer there, on behalf of himself and his collaborator, Robert Kirshner.

Usually spectra obtained with IUE are of faint stars, with exposure times of minutes to hours. This spectrum was over-exposed at 15 s and, for the bright supernova, should have been just a 3 s snap.

More than 50 spectra of the supernova were obtained over the next week, from GSFC and Vilspa.

The first photograph

It was Shelton's photograph on which the supernova was first noticed. The supernova was actually first recorded photographically by McNaught.

Since September 1986, using 35 mm cameras and telephoto lenses, McNaught and Gordon Garradd in Tamworth, NSW, have photographed the LMC regularly, in order to discover novae. Usually they check their exposures within 24 hours of making them. McNaught had made two exposures of the LMC on February 23 between 11:00 and 12:00 UT, and developed and mounted them within a few hours.

'For reasons which I will never understand,' he records (McNaught 1987a), 'I did not search them, despite telling myself several times that day that I

Fig. 5. McNaught took three pictures of the LMC before, during and after the outburst of SN 1987A, with an f/2.4 85 mm focal length lens on a conventional 35 mm camera loaded with Tri X film. The LMC shows as a bar of stars with bright nebulae in two rudimentary spiral arms to the lower right and upper left. The bright spot in the upper left spiral arm is 30 Doradus and the supernova grows in brightness just below it.

On February 23 McNaught caught the supernova at sixth magnitude just south-east of the bright nebula 30 Doradus. This is the first picture of the supernova. On February 24 the supernova was magnitude 4. © R.H. McNaught.

(*a*) 1987 February 22 11 : 14 UT, exposure 5 min.

(*b*) 1987 February 23 10 : 42 UT, exposure 2 min.

(*c*) 1987 February 24 10 : 47 UT, exposure 5 min

should.' Distracted from the photographs by his long term aim to discover novae and distinguish them from known variable stars, McNaught started to key in the coordinates of all LMC variables to allow computer identification of suspects instead of laboriously searching through catalogues. On the photographs, at magnitude 6.0, was the supernova, unseen until McNaught looked, following Jones' report.

Missing the discovery of the supernova was water under the bridge for McNaught. No further time should be lost. After learning of the supernova's discovery, McNaught took a quick photograph with the RGO's Hewitt Camera, which is normally used for tracking satellites by photographing them against a background of stars. Then he measured the supernova's position.

McNaught noticed that the supernova was coincident with the position of a blue star which had been visible on old plates. Could this be the progenitor star? – the star which had exploded? No-one had identified the seat of a supernova explosion before; had it been giving clues in the past that it was going to blow up?

Some stars are variable in brightness: it was conceivable that a star which was about to blow up would show some sign of instability by varying. Mati Morel, an amateur who charts variable stars for the Royal Astronomical Society of New Zealand, took an interest in this point. Was anything known about McNaught's blue star which implied that it was a variable? First, Morel noticed that the star did indeed have a name – it was Sanduleak −69 202. But it was not variable: checking back on previous records of the star, Morel and McNaught could find no indication of its variability over the last century. There was no indication in its light output that it was about to explode, and in fact there are good theoretical reasons why there would not be (Chapter 14).

McNaught communicated the identification of the progenitor star and its lack of variability immediately to the Central Bureau.

Africa sees the supernova

Night-time was still travelling around the world towards Africa, while astronomers were reading the Central Bureau's astrogram.

Darkness reached Zimbabwe in southern Africa where Colin Henshaw went out that evening, as usual, to make observations of bright variable stars. With the naked eye he noticed what he at first interpreted as an unusually conspicuous visibility of the Tarantula Nebula in the LMC. Checking with binoculars he found a fourth magnitude star south-west of the nebula, where he was sure there was none before; he realised that he had found a supernova.

Table 1. *The first day, 1987 February 24*

01:20	Shelton	Las Campanas	3 h discovery photograph begun
03:00	Duhalde	Las Campanas	Suspected visual sighting
04:20	Shelton	Las Campanas	Exposure terminated
05:40	Shelton	Las Campanas	Discovers supernova on plate and in sky
07:55	Duhalde	Las Campanas	Magnitude[a] $v = 4.5$
09:00	Jones	Nelson NZ	Discovers supernova $v = 5.6–7.0$
10:55	Jones	Nelson NZ	$v = 5.1$
10:54	Moreno	Auckland NZ	$V = 4.81$
11:20	McNaught	Coonabarabran	$v = 4.8$
15:00	Green	Cambridge, Mass.	Issues IAU telegram
15:14	McNaught	Coonabarabran	$v = 4.4$
17:12	McNaught	Coonabarabran	$v = 4.4$
19:00	Kirshner	GSFC	IUE satellite on supernova
19:12		Cape Town	$y = 4.72$
19:16	Henshaw	Zimbabwe	$v = 4.6$
19:39		Sutherland SA	$V = 4.626$
19:56	Henshaw	Zimbabwe	$v = 4.4$
20:04	Kirshner	GSFC	First ultraviolet spectrum
20:09	Henshaw	Zimbabwe	$m_{vis} = 4.6$
21:08	Henshaw	Zimbabwe	$v = 4.1$
21:40	Menzies	Sutherland SA	First optical spectrum
22:47	Henshaw	Zimbabwe	$v = 4.0$
01:30	Marsden	Cambridge, Mass.	Issues second IAU telegram

[a] Stellar magnitudes are a way of describing the brightness of stars. Stars of the first magnitude are the brightest ones, stars of the sixth magnitude are the faintest ones which can be seen with the unaided eye, so the smaller the number of a stellar magnitude, the brighter the supernova. Of course, there is considerable uncertainty in estimates by eye. In this table, a magnitude quoted to 0.1 denotes a magnitude estimated by eye. A magnitude quoted to 0.01 or better denotes one estimated with a photometer, or by accurate measurement of a photograph. The lower case v is used to represent a visual magnitude determined by eye, upper case V represents a so-called visual magnitude determined with a photometer, y represents a yellow magnitude determined with a photometer, m_{pg} represents a magnitude determined photographically, and m_{vis} represents a so-called visual magnitude determined photographically. Whilst in principle all these methods would yield the same value for the supernova if measured at the same moment, in practice they might be different.

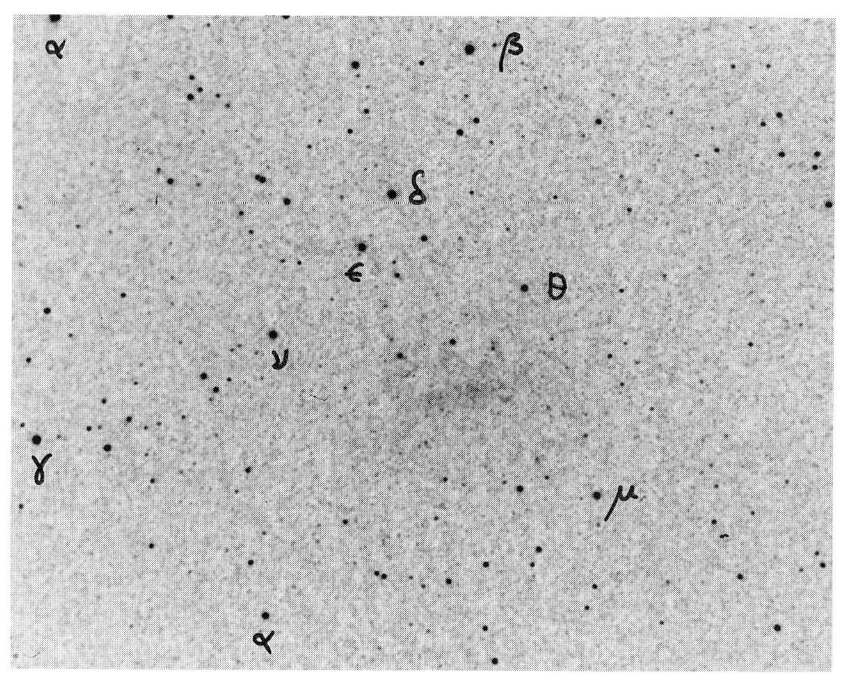

(*a*) Before the supernova

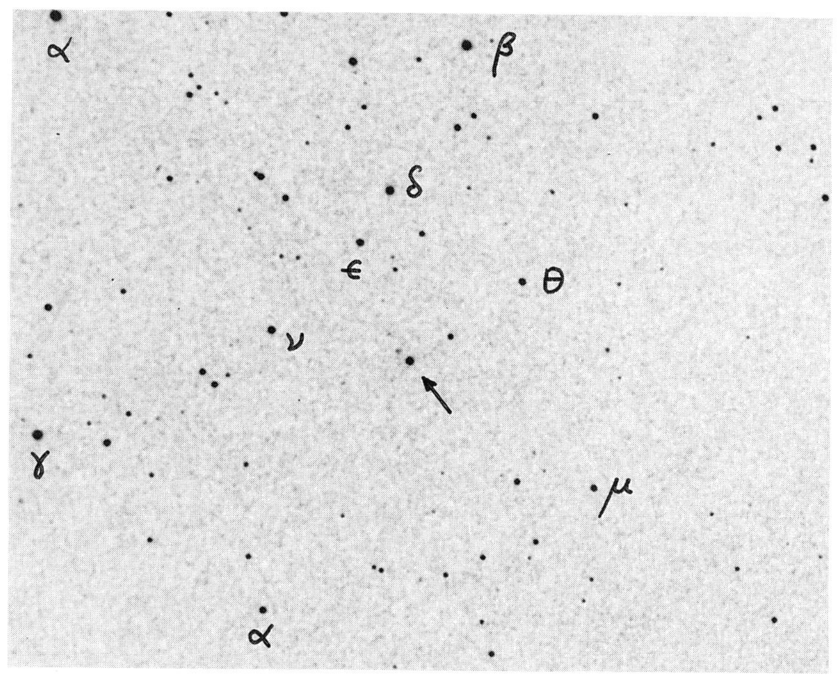

(*b*) 1987 February 24 20 : 09 UT

Fig. 6. Henshaw took an unguided 20 s exposure of the supernova (arrowed in (*b*)), with his camera loaded with 1000 ASA film, on 1987 February 24 at 20:09 UT. Here the picture is compared with a somewhat longer exposure which he took before the supernova appeared: the earlier picture (*a*) records the LMC as a grey bar in these negative prints.

The LMC itself is in the contellation Dorado, which stretches to the upper right (north-west) of the pictures. The constellation Pictor is at the top, Volans to the left, Mensa to the bottom, Hydra to the lower right and Reticulum to the right. The stars named have the following magnitudes, with the brighter stars (larger black images) to the top of the list:

Alpha Pic	**3.27**
Beta Dor	**3.76**
Gamma Vol	**3.78**
Delta Dor	**4.35**
Theta Dor	**4.83**
Nu Dor	**5.06**
Alpha Men	**5.09**
Epsilon Dor	**5.11**
Mu Men	**5.54**

The supernova's image is just smaller than Delta Doradus, but larger than Theta or Nu Doradus. By interpolation of measured images, its magnitude was 4.6.

After preparing his charts he estimated its brightness at 19:16 UT at magnitude 4.6, and took photographs (Figure 6(*b*). I calibrated Henshaw's photograph taken at 20:09 UT: it gave $m_{vis} = 4.6$. The supernova brightened as Henshaw watched: it had reached magnitude 4.0 by 22:47 UT. Henshaw failed to make contact with a friend with a telex machine and so could not notify astronomers in other countries.

Meanwhile, night had fallen at the Cape, in South Africa. In the Karoo Desert, at the Sutherland station of the South African Astronomical Observatory (SAAO) John Menzies was observing with the 1.9 m Radcliffe Telescope. It was fitted with a spectrograph. Nearby, the 0.5 m reflector was in operation, equipped with a photometer. In Cape Town itself the Cape astronomers opened up the SAAO's 0.46 m telescope, with its photometer. Alerted by Brian Warner, a Cape Town astronomer who was at the time in Texas and had heard of the supernova by the 'jungle drums', the South African astronomers made their first observations.

Menzies obtained the first optical spectrum of the supernova on February 24 21:40 UT. The supernova was very blue and showed extremely broad absorption lines. One of them was near 6150 Å. Menzies, by his own reckoning not a supernova expert, telephoned back to Warner in Texas, and described the appearance of the spectrum to him. In a lengthy conversation they concluded that the absorption feature came from ionised silicon. Supernovae may be of two types, and a silicon feature in the spectrum would be indicative of Type I. The South African astronomers concluded that SN 1987A was maybe a Type I supernova and told the IAU Central Bureau for Astronomical Telegrams so.

The feature was in fact an absorption due to hydrogen at an amazing blue shift and the supernova was Type II.

The Central Bureau's director, Brian Marsden, issued a second telegram at 01:30 UT.

> SHELTON DUHALDE JONES 1987A LMC SUPERNOVA MCNAUGHT
> 19502 70224 05355 02216 91759 25045 14101 24375
> BRIGHTENING ONE MAGNITUDE DAILY
> BLUE NONVARIABLE MAGNITUDE TWELVE OBJECT WITHIN
> ONE ARCSEC THROUGH 70222
> MENZIES SPECTRUM SHOWS MAYBE TYPE ONE
> REF EARLIER MSG MCNAUGHT NOT CODISCOVERER
> MARSDEN 5FEB25/0130Z

This telegram carried McNaught's report on the supernova:

> 19502: McNaught was supplying an accurate position for the supernova, expressed, not in telescope coordinates, but in coordinates of 1950 ('2' at the end of the first group means the position is 'accurate').
> 70224: As in the previous telegram, this is the date when the observations were made: [198]7 02 24.
> 05355 02216 91759 25045: Since this is an 'accurate' position, the positional data has more digits than before and runs into the magnitude group. The supernova is at: RA05h 35min 50.22s, Dec. $-69°$ $17'59.2''$, and the visual magnitude is noted at 4.5.
> The checksums are: $19502+70224+05355+02216+91759+25045=21401$, with the leading digit 2 dropped, and $05355+02216+91759+25045=124375$, with the leading 1 dropped.

McNaught's discovery noted that the supernova is coincident with a blue star, non-variable up to February 22 of the current year. The telegram distributes Menzies' incorrect conclusion about the supernova's type and Marsden corrected the implication in the previous message that McNaught was a co-discoverer.

Table 1 summarises the events of the first day of the supernova in the LMC.

Night returns to Chile

Thirty kilometres from Las Campanas Observatory, from where the supernova was discovered, lies the mountain top of La Silla, where the

Table 2. *The onset of the explosion*

Feb 22	11:58	Garratt	Tamworth NSW	No supernova on LMC patrol photograph
Feb 23	01:11	Madsen	La Silla	No supernova on colour exposure
	01:55	Shelton	Las Campanas	No supernova on plate
	09:22	Jones	Nelson NZ	Did not notice supernova with small telescope so it was fainter than $v = 7.5$
	10:39	McNaught	Connabarabran	Supernova magnitude $m_{\rm vis} = 6.4$
	11:17	McNaught	Coonabarabran	Supernova magnitude $m_{\rm vis} = 6.0$
	12:58	Thomas and Ryder	Dunedin	Supernova on a photograph $m_{\rm vis} < 7.5$
	14:21	Zoltowski	Woomera	Supernova magnitude $m_{pg} = 6.1$

European Southern Observatory (ESO) is built. The two observatories are within sight of each other (Plate 2).

The first that ESO knew of the supernova was upon receipt of the IAU telegram. The information had echoed from mountain top to mountain top via Cambridge, Massachusetts. Acting-Director Hans-Emil Schuster walked into the dining room at tea time – time for the La Silla astronomers' breakfast, since they had slept from dawn after the previous night. Schuster told the assembled astronomers of the LMC supernova. Not knowing the observatory from which Shelton's report emanated – it was not mentioned in the first IAU telegram – Schuster at first appeared to be sceptical about the event. But, as darkness fell, a few hours later, the ESO astronomers confirmed the report and the first observations of the supernova were beginning to be made on the La Silla mountain (Hanuschik, Thimm and Dachs 1988) by the considerable array of one of the world's largest collections of astronomical telescopes.

At the Cerro Tololo Interamerican Observatory (CTIO) they also received the IAU telegram. That afternoon, in a frantic effort, technical staff pressed rapidly into service a small, idle 0.4 m (16 in) telescope (Maran 1988). It made its first measurements of the supernova's brightness that night.

When did the star explode?

McNaught's photographs of February 23 were the earliest of the supernova in explosion. Others had been taken soon afterwards (Table 2). In Dunedin, NZ, M. Thomas and S. Ryder, two astonomers testing a new telescope recorded the supernova on the test plate. Frank Zoltowski, an amateur in

Woomera, Australia, recorded it on a photograph of the LMC which he was taking to test sensitised film.

But to pin down the onset of the supernova explosion it is necessary to know not only when the supernova was first recorded, but also when it was last not recorded.

Photographs taken just before McNaught's prediscovery photographs, but showing the LMC with no sign of the supernova, are also listed in Table 2. The most significant gap lies between Shelton's photograph which he exposed in Chile the day before discovering the supernova, and McNaught's uninspected prediscovery photograph taken from Australia.

As Marsden dryly commented: 'There is obviously much interest in prediscovery photographs that may have been taken of the supernova field between February 23.1 and 23.4 UT, although it seems unlikely that there would have been many observers between Chile and New Zealand.' No photographs of the appropriate area, taken in this time interval, have turned up; there are, of course, few astronomers living on Pacific Ocean islands.

Ginga

By February 25 the world's astronomers had had time to respond to Shelton's discovery and the telegrams of the IAU's Central Bureau.

In Tokyo, Japanese astronomers of the Institute of Space and Aeronautical Science and their University of Leicester and Rutherford-Appleton Laboratory collaborators had been, for two weeks, poring over the test data from the newly launched artificial satellite *Astro-C* when they heard the news.

Launched on February 5, the satellite had been renamed 'Ginga' meaning 'Galaxy'. The satellite, designed for X-ray astronomy, contained equipment intended specially to examine the whole sky to monitor X-ray stars for variability. A Large Area Counter (LAC) was another detector on board which was intended to scan areas of the sky for good intensity and spectral information of known X-ray sources.

The LAC had not in fact been completely activated when the news came through of the supernova (Turner 1988). It was being slowly brought up to operation, with the space engineers learning from a bad experience in a previous experiment on the satellite EXOSAT.† The idea in Ginga was to be very careful in checking out the detectors.

† The high voltage supply which powered some detectors on EXOSAT broke them down and thenceforth they discharged continuously if they were operated at full voltage. It had been necessary for the rest of their lifetime to run the detectors at half their gain, thus losing sensitivity.

Furthermore, the satellite was in direct view of the ground station for only a short time each day. This was the only period when it could be monitored from moment to moment to be confident that it was in no danger. When the satellite was out of direct view, it relied on automatic devices for protection.

For example, there was a danger every orbit when the satellite passed through the area of high radiation known as the South Atlantic Anomaly. The Anomaly is a magnetic bottle in the Earth's magnetic field and traps cosmic rays from the Sun. The danger to Ginga was that the high radiation level could cause its high voltages to discharge, damaging the equipment as in the EXOSAT case. Automatic shutdown devices were programmed to protect Ginga's equipment as it passed through the Anomaly, but had not been checked. If the equipment was brought into operation immediately to observe the supernova without the full checks being carried out, there was a small risk to the future operation of the satellite.

The astronomers pressed the space engineers to curtail the checks in order to take advantage of the unique observing opportunities given by the LMC supernova.

The scientific benefits of immediate observations of the supernova were weighed against the risks of curtailing the test programme, and the operations team decided to take the gamble. Parts of the test programme were postponed in order to get data on the LMC supernova – it was judged very important to observe the supernova immediately.

The gamble paid off. All the equipment worked properly and Ginga survived intact. The observations of the supernova were correctly carried out. Although no signal from the supernova was detected for several months, this early null result, followed by a later detection, turned out to be a crucial piece of data. The time that X-rays became visible was a test of theoreticians' understanding of the exploding supernova (Chapter 10).

The Ginga operations team decided to continue with a new programme to monitor the area of the supernova, to see when it appeared. They fitted the test programme, and their originally planned programme of non-supernova observations, around the new supernova programme.

Radio observatories

Australian radio astronomers were immediately successful in detecting the supernova. The Fleurs Radio Observatory and the Molongolo Observatory in Sydney were the first radio telescopes to pick it up at 10:00 UT on February 25, at radio frequencies of 1.4 GHz (21 cm wavelength) and 0.8 GHz respectively.

Fig. 7. A map (by Turtle *et al.* 1987) of the radio emission from the LMC was made at a frequency of 843 MHz with the Molongolo Observatory Synthesis Telescope, after the supernova had appeared. It shows the brighter radio sources as the higher hills. The 30 Doradus Nebula is the highest peak in the picture. Some of the other peaks are radio sources in the LMC and others are background galaxies and quasars. The supernova is the peak arrowed in the right of the picture obtained 1987 February 26. This radio picture is directly comparable with the optical picture in Figure 4; the LMC nebulae are visible in both pictures but the radio quasars are not discernible in the optical pictures and the optical stars have no bright radio counterparts.

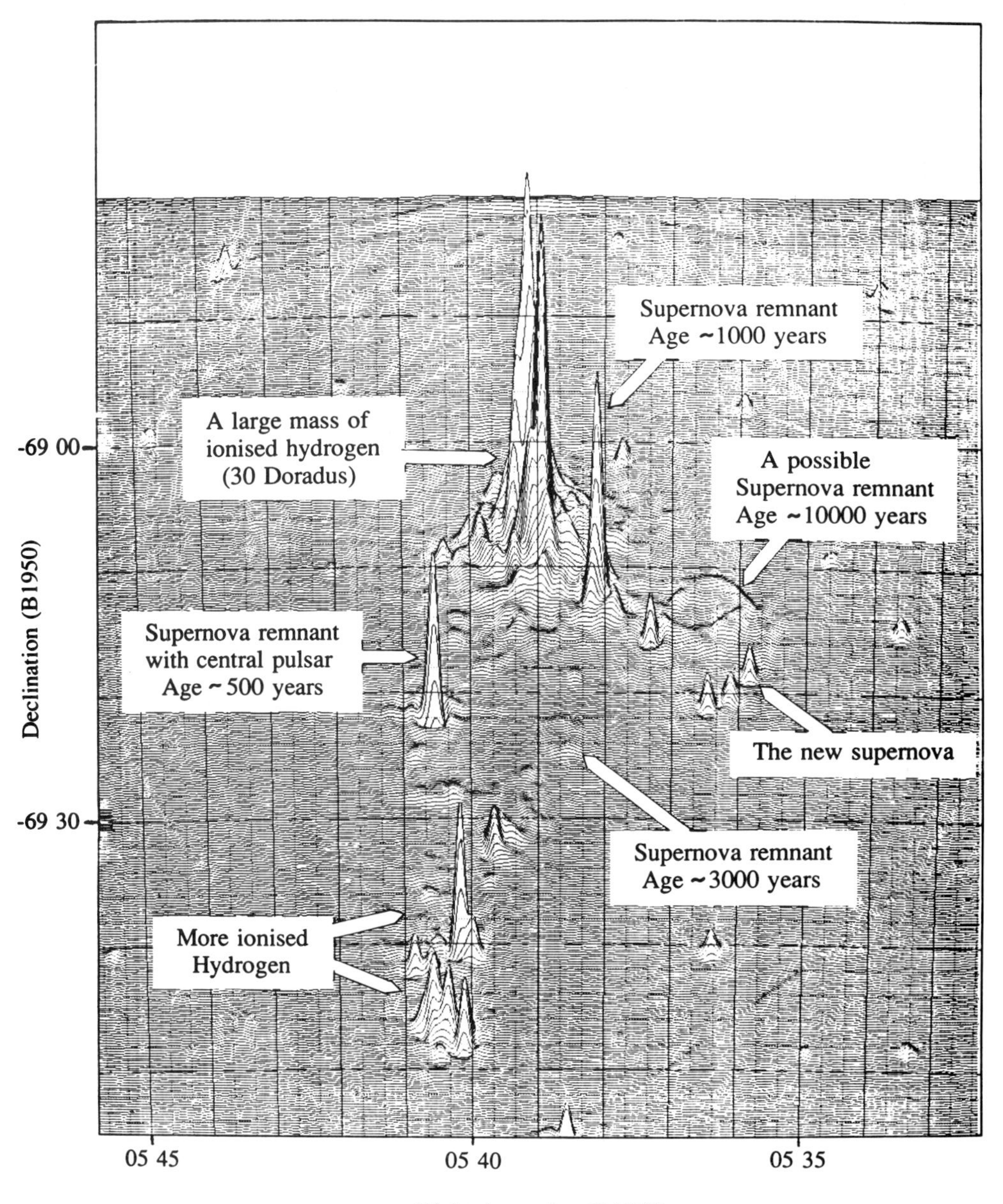

This early radio detection of the supernova was very surprising – it was the first time that any supernova had been detected by radio telescopes so soon after its optical discovery.

The next day it was detected by the Parkes 64 m telescope of the Commonwealth Scientific and Industrial Research Organisation (CSIRO) Division of Radiophysics and by one of the 34 m dishes of NASA's Deep Space Communication Complex at Tidbinbilla. Normally the Tidbinbilla dish is used for tracking satellites and at the time was occupied with the Voyager spacecraft. Fortunately Voyager was in the opposite direction to the LMC and the dish could be used for both studies without difficulties about scheduling priorities.

But the most dramatic detection of the supernova had been made, not by radio observatories, nor by the optical observatories on the distant mountain tops of the world, but by a new kind of observatory buried deep *within* mountains or in mines below the Earth's surface. Astronomers at the world's neutrino observatories were searching back through their records for the slimmest evidence of the supernova outburst (Chapters 7 and 8).

3
The astronomers react

Making and unmaking plans

A big modern telescope costs approximately £10 million to create and £1 million per year to run. Each year there are 365 nights on which it can work, but astronomers apply to use it for 1000–2000 nights per year. In other words, a large telescope is oversubscribed three or four times, and the pressure to use it is great. A telescope is therefore scheduled for use every night, long in advance. When the supernova occurred, unexpectedly, astronomers had to work outside the normal rules to seize the unique opportunity to study it.

Some observatories set aside a few nights in the year to be able to respond to new discoveries, or to make observations on behalf of people who want a brief look at something. These are called 'Director's nights' or 'service nights' respectively. It was relatively straightforward to use these for the study of the supernova as they turned up in the schedule. But they were not enough and it was necessary to override telescope schedules.

A typical schedule lists, night by night, the names of the group of astronomers who have been given the right to use the telescope for two or three nights, and the instrument which they will use. Basically the telescope gathers the starlight and feeds it to an instrument where the light is analysed in a particular way to yield the information required. Each instrument is individually mounted rigidly onto the telescope and instruments have to be swapped on and off the telescope according to the analysis needed – there are very few telescopes on which several instruments can be mounted at once. It is often not an easy job to take off from or fit to a telescope a delicate instrument weighing tons.

What is more, once the instrument is fitted it has to be set up – there will be several adjustments (mechanical, optical, electronic) which must be carefully tuned. It may need a time to settle down to full accuracy after being activated. Setting up an astronomical instrument is often not a quick task. The telescope schedule will have been organised to facilitate instrument changes and their setting up, including the all-important requirement that the right engineers and technicians – the ones who know the instrument and how to get it working well – must be on hand to carry out the work.

Even when an instrument has been changed to something suitable for a supernova study, there will be another element in the schedule to adjust – and this element, being human, will have the biggest inertia of all!

The group of astronomers scheduled to use the telescope will have been making plans for several months for how they will use the observing run of two or three nights; it is likely that the study which they intend to execute will be part of a long term plan. It may be that the observations are to be used for a student's PhD thesis: he or she has three years in which to decide what to do, make the observations and think about what they mean. If the observations do not get done, the student might be in deep trouble, because it might be necessary to wait for another year for a second try.

Thus a group of astronomers who have travelled from their European or North American city half way around the world to a distant mountain top in Australia or Chile will be reluctant to give up the night which they have come to think of as 'theirs' just because something has happened which is irrelevant to their work, however important in principle its study. As one observatory director put it: 'Everyone agreed that this supernova was so important that, clearly, we ought to devote a lot of time to it; but everyone would much rather have somebody else give up his time' (Cannon 1987).

So the first problem in organising the study of the supernova with optical telescopes was having to override the scheduled observers and divert them from their intended study.

At ESO in Chile, the Director-General, Lew Woltjer, was making one of his periodic visits from the ESO headquarters at Garching near Munich, Germany. On the mountain top at the same time was John Danziger, a senior ESO astronomer, observing with the 3.6 m telescope. They encouraged all the astronomers present on the mountain to start observing the supernova from the night of February 24/5 onwards. This was truly a unique occasion on La Silla, with a diverse community of European scientists involved (and self-consciously so) in a common endeavour. Indeed at one point in the first night of observations, the control room of the 3.6 m telescope was so jammed with bystanders that it was difficult to observe efficiently. 'All observers at La Silla have switched over to the supernova with much enthusiasm,' wrote Woltjer (1987). 'While for their coinvestigators it may be disappointing not to receive the data from their planned programmes, we expect that they will understand that it was not possible to continue business as usual in the light of this event.' The excitement and pleasure that running the new form of the business at ESO created was heightened by setting up daily Astronomers' Teas†. Everybody reported the last night's results, sketching on graph paper with coloured pencils; a great sense of community was created, as the astronomers focused on their common point of study.

At the SAAO, the Director, Michael Feast 'gently persuaded' astronomers to give up their telescope time for the supernova observations (Menzies *et al.* 1987).

At the AAO (Plate 3), the newly-appointed Director, Russell Cannon was flung headlong into the maelstrom; telephoning the chairmen of the two independent bodies, Australian and British, which give out time on the AAT, he agreed that he would make available for the supernova an hour of every night on the telescope, whatever instrumentation was on the telescope. From time to time he would make up to half a given night available for specialist investigations. All the time which he had available at his discretion (10% of the total) would, in the immediate future, be dedicated to the supernova (Couch, 1988).

Those astronomers who suffered from this override were offered some compensation of access to the AAT in the Director's discretionary time at a later date, when it was not so important to monitor the supernova so frequently. All of them agreed, some with more grace than others, to accept the override.

Supernova-mania

The pressure on observatory directors is well expressed by David Allen's account from the Anglo-Australian Observatory (AAO): 'From the moment we heard [of the discovery of the supernova], life has not been the same. It will be difficult for readers at northern institutions to appreciate the impact of the supernova. We have shuffled the schedule, taken over service time, overridden observers; we have used up to three instruments per night, even arranging major changes part way through the night; we have cobbled together an instrumental configuration never before used, and we are building an ultrahigh resolution system for the coudé to take advantage of the SN to probe the

† Schuster's arrangement to provide sticky cakes was sometimes greeted with blunt comments from astronomers who had just got up and faced the cakes at what they considered breakfast.

interstellar medium. None of us being supernova pundits, we have agonised for many man-days to make many decisions embodied in this list of activities, and we have done so under a barrage of public and media interest exceeding in its intensity the height of Halleymania' (Allen 1987).

Making telescopes smaller

The second problem in the study of SN 1987A was its brightness – it was too bright! Astronomers strive with the instruments which they use to see and measure the faintest possible stars. They build bigger and bigger telescopes for this purpose, bringing more, and more distant, stars within their knowledge. The instruments with which they analyse the light from faint stars are matched to the power of the telescopes – in much the same way that the individual components of a hi-fi sound system (for example, turn-table and amplifier) are matched in performance, facilities and cost. This can mean that astronomers simply cannot use a given astronomical instrument and telescope on bright stars.

The supernova was discovered a few days before the New Moon; the sky is at its darkest at this time of the month, and most telescopes and instruments were, at the time, set up to study the faintest stars, uninhibited by moonlight.

At Las Campanas, for example, when the supernova was first discovered, one telescope was equipped with a CCD (Charge Coupled Device). This is an array of diodes which are very sensitive to light which they convert to electric charge and store. The charge can then be read from the diodes and be used to determine the brightness of a star. The problem is that a CCD can only store a small amount of charge; if more than the maximum amount falls on a diode then it saturates, and however much more light falls on it, the diode ignores it. To control the amount of light which is allowed to fall on the CCD, a shutter is placed over it. The shutter will allow light through for a certain time – usually a long time, since the stars are usually faint and the exposure may need to build for several minutes or even hours. There is a limit to the speed at which a mechanical shutter can sweep across a given sized CCD chip, and still maintain a uniform exposure. The Las Campanas CCD would not operate in fact for less than five seconds. Even in that short time the CCD would saturate when exposed to the supernova, and it was impossible accurately to measure the supernova's brightness with it. The observers could not, on that first night, find a way of attenuating the supernova's light.

At La Silla, one of the most suitable spectrographs to study the supernova was attached to the 3.6 m telescope. The scheduled programme (Fosbury 1988) was to study 18–20th magnitude radio galaxies. The astronomers had to change to study SN 1987A, which was a million times brighter. The spectrograph had a CCD as a detector, with the same limitations as the Las Campanas CCD, and so the astronomers fiddled with the electronic gain of the CCD, making it less efficient. They drifted the supernova along the slit of the spectrograph, spreading its light over more of the diodes in the CCD and avoiding saturation that way. Because of thc amount of light which the CCD stored the spectra were of very high accuracy – probably the highest accuracy of the early low dispersion spectra.

At the AAO they also began to study the spectrum of the supernova, recording it with CCDs and with the Image Photon Counting System (IPCS). It was necessary to attenuate the light from the supernova by a factor of 1000 to avoid overloading the detectors. This fact severely tested the goodwill and patience of astronomers whose use of the large telescope had been overridden for supernova studies; the 3.9 m AAT was being made into the equivalent of an amateur-sized 12 cm telescope – admittedly with a £1 million set of instruments attached!

A programme was begun on the AAT to measure the brightness of the supernova to see if there were fast pulsations from any pulsar which might have been created, using an auxiliary photometer looking through the large telescope. The photometer was attenuated with a similarly powerful filter, but because this was uneconomic, the programme was quickly converted to use an off-the-shelf amateur reflecting telescope, which stood on the ground floor of the observatory peering out of an open door. The small telescope fed the signal upstairs to the fast-photometry computer at a time when it otherwise was idle, and another computer was in operation attached to the main telescope. Here was a case where an inexpensive telescope and an expensive instrument were deliberately brought together in an apparent mismatch because of the unique circumstances of the supernova.

The intensity of the bright signal from the supernova was also a problem to South African astronomers. Attached to their telescopes were photometers used to measure the brightness of stars. They were pulse-counting: each photon was converted into an electronic pulse and individually counted. These pulses have a certain duration set by the electronic characteristics of the photometers, and if too many pulses are generated they pile up into a continuous train in which individual pulses cannot be distinguished. Fortunately, since the SAAO astronomers do occasionally need to measure bright stars, they had immediately available filters which could be put into the light beam from the supernova to dim its light to a countable rate. Then it was a matter of checking the calibration of the filters so that their attenuation rate could be accounted for.

Even so, as the supernova brightened to magnitude 2.8 a mask had to be constructed for use on the 50 cm telescope, comprising two 10 cm apertures placed over the end of the telescope and making it effectively a 15 cm telescope – an amateur-sized 6 in (Catchpole *et al.* 1987).

The same problem afflicted astronomers at the CTIO. They began a series of observations of the supernova's brightness with a small 0.4 m (16 in) telescope (Maran 1988). The telescope contained a sensitive photomultiplier (an RCA 31034 with a gallium-arsenide photocathode). 'In order to avoid damaging the detector,' wrote Hamuy *et al.* (1988), 'it was necessary to reduce the aperture of the telescope with a Hartmann mask, which is a piece of aluminium perforated with 26 one inch holes that covers the entrance pupil of the telescope. As the supernova brightened the telescope's effective area was progressively decreased until at maximum light it was only [about] four square inches.' They did this by sticking black tape over more and more holes until they were down to just four open ones.

It is not usual for astronomers to throw light away, making their telescopes smaller! – mostly the cry is for larger and larger telescopes.

The Half-Baked Spectrograph

The brightness of the supernova threw up a weakness in the equipment available at the AAO. There was no spectrograph which could study the supernova's spectrum at the really high resolution which it deserved. In fact, in the original plans for the telescope, such a spectrograph had been envisaged; it had been dropped because of pressure on the budget. In spite of energetic, articulate, forceful and persistent lobbying by a high resolution spectroscopist, Roger Griffin of Cambridge University, only in 1985 had a firm decision been made to build a facility for high resolution spectroscopy at the AAT – even then, Griffin was not exactly enchanted with the proposals made. When the supernova exploded, the new spectrograph was under construction in Britain, and not at an advanced enough stage to be shipped to and used in Australia.

However, the Chief Engineer at the AAO, Peter Gillingham, realised that, if he put his family responsibilities in second place (Peter's wife, Mary, is resigned to this, having suffered while her husband commissioned the AAT in 1974 – and on other occasions), he could quickly make a suitable spectrograph using a diffraction grating already purchased for the spectrograph under construction. He resolved to do so after having to placate a scheduled astronomer, displaced for a couple of hours while observations were made which, however necessary, did not use the full capacity of the large telescope. Given the likely fast fading of the supernova, the spectrograph had to be completed to an urgent schedule – no-one forecast that this supernova was going to last at its maximum brightness for so long.

A spectrograph's resolution depends on the diameter of the beam which it uses – high resolution spectrographs need large gratings. As a consequence they need a large collimator lens to feed the light beam to the grating and a similarly large camera lens to collect the light beam off the grating and focus onto a detector the spectrum which it produces.

An AAT night assistant, Steve Lee, a keen amateur astronomer, makes his own mirrors for astronomical telescopes. He used his amateur's equipment on a blank which Gillingham scrounged from some friends at the Defence Research Centre, South Australia, to make a 7 m focal length lens, 21 cm in diameter.

Initially, it seemed desirable to use fused silica for this lens, so that it passed ultraviolet light. Gillingham had been able to borrow such a blank without

Fig. 8 (*a*). A spectograph contains a diffraction grating to disperse the light into the colours of the rainbow. The spectrum is focused on a detector by a camera lens. The diverging beam from the telescope is rendered parallel by a collimator lens which feeds the grating.
(*b*) If a spectrum of light between red and blue is to be spread over a larger distance to improve the resolution of a spectrograph, the diffraction grating in the spectrograph has to be bigger. So too, do the camera lens and the collimator lens.
(*c*) If the grating is made reflective and tilted, the collimator lens can double up as the camera lens.

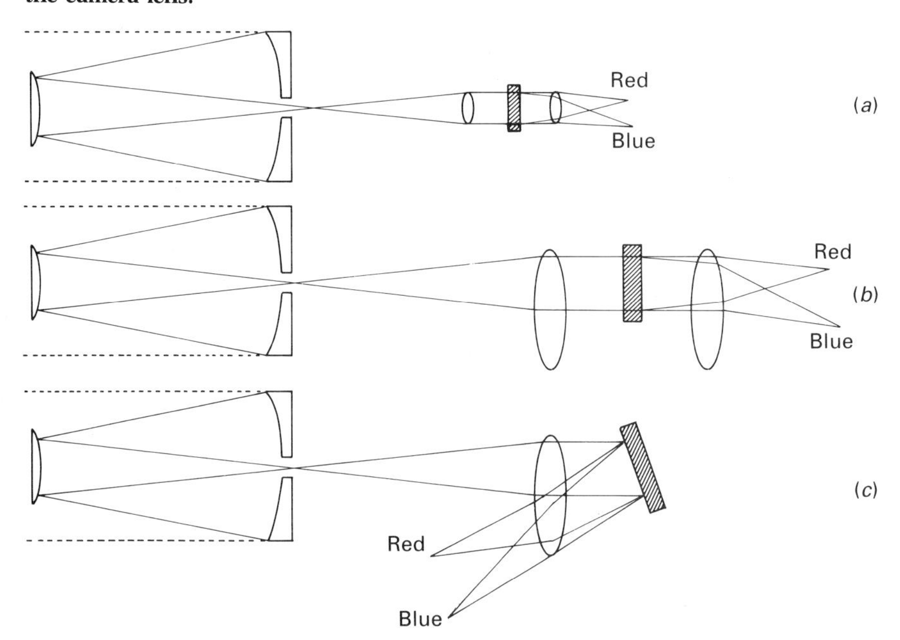

delay but its quality was uncertain. Eventually they received word from its manufacturers that the silica was not sufficiently homogeneous to give well-focused images (the refractive index was different in the various lumps of silica fused together in the blank). By that time Gillingham and Lee had found this out the hard way: they had completed the lens.

They located and bought a second blank – a top quality one of glass: by this time they had realised that the loss of ultraviolet transmission in glass did not matter, since the supernova had faded very rapidly in the ultraviolet and they would now never be able to use the special ultraviolet transmission of the silica lens. Lee ground and polished this blank into another lens, walking several thousand times around the 44 gallon drum which he used as a table.

This one lens was used in 'double pass' – the light from the telescope was sent through the lens to collimate it to a parallel beam, and collected by the same lens after reflection from the diffraction grating to be focused by the camera on the detector. The design economically and speedily fulfilled the normal requirement of two large lenses to match the large grating by using the same lens twice. Other equipment for the optical beam was borrowed from other telescopes in Australia – a low dispersion prism was from the Swedish Uppsala Schmidt Telescope, for example. The whole arrangement was built into a wooden frame, very much a temporary arrangement and one of which Griffin would not have approved, but it worked!

In his description of the spectrograph, Gillingham gives special thanks to an unnamed astronomer. This man responded to a request for information, before the project was started, only by condemning the project to others as 'half-baked'. 'Learning in a roundabout way of his response was a very effective spur to proving the scheme would work,' wrote Gillingham (1987), 'and provided the instrument with its working title: the Half-Baked Spectrograph.'

This string and sticky tape approach to high resolution spectroscopy was not necessary at the well-funded ESO. A high dispersion spectrograph called the Coudé Echelle Spectrometer (CES) is routinely available in the dome of the 3.6 m telescope. It is fed by the coudé beam of the telescope. Adjacent to the large telescope is a second, 1.4 m telescope which also feeds this spectrograph, its light beam entering the other telescope's building along an 11 m umbilical steel tube. The so-called Coudé Auxiliary Telescope (CAT) stands on a 24 m high pillar alongside the larger telescope ready to feed light into the CES at any time, independently of what the larger telescope is doing. Since the CES can be fed by the large telescope if the astronomer is observing faint stars or the small telescope if he is observing brighter ones it is always available no matter how bright the stars are.

Fig. 9(*a*) Essential components of the AAT's Half-Baked Spectrograph were the echelle diffraction grating and the new collimator and camera lens. Because the grating was used at a high order number, the individual orders of the spectra would overlap unless the prism was used to separate them by prismatic dispersion. (*b*) The business end of the AAT's Half-Baked Spectrograph had two detectors ready to fire up to record the supernova: the large light-coloured cylinder contains the image intensifier of the Image Photon Counting System (IPCS) and the smaller grey-coloured cylinder at the top is a cryostat or cold chamber which contains a CCD. The spectrograph lies beyond the black curtain and light from the supernova enters the spectrograph from behind the reader's left shoulder at the entrance slit at the centre of the photograph. Photo by Duncan Waldron.

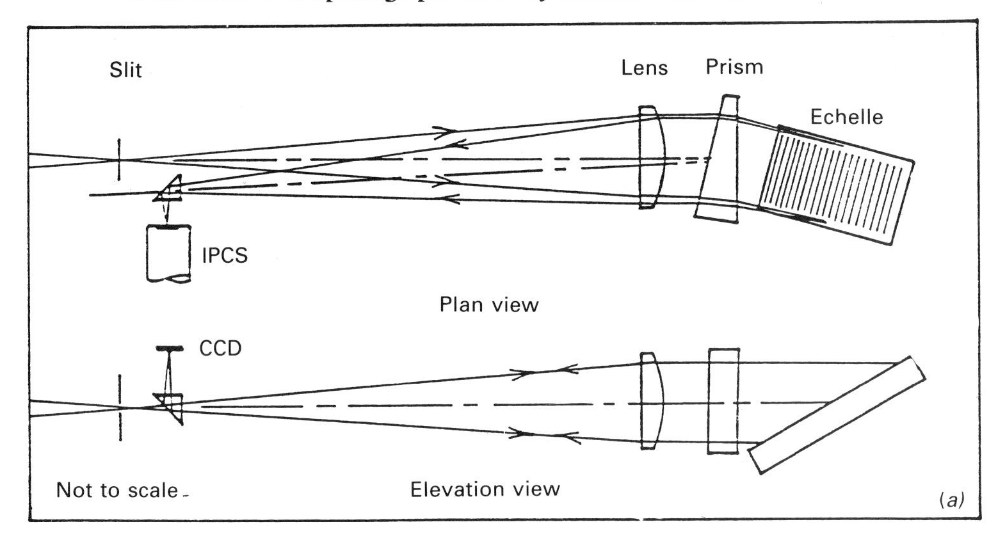

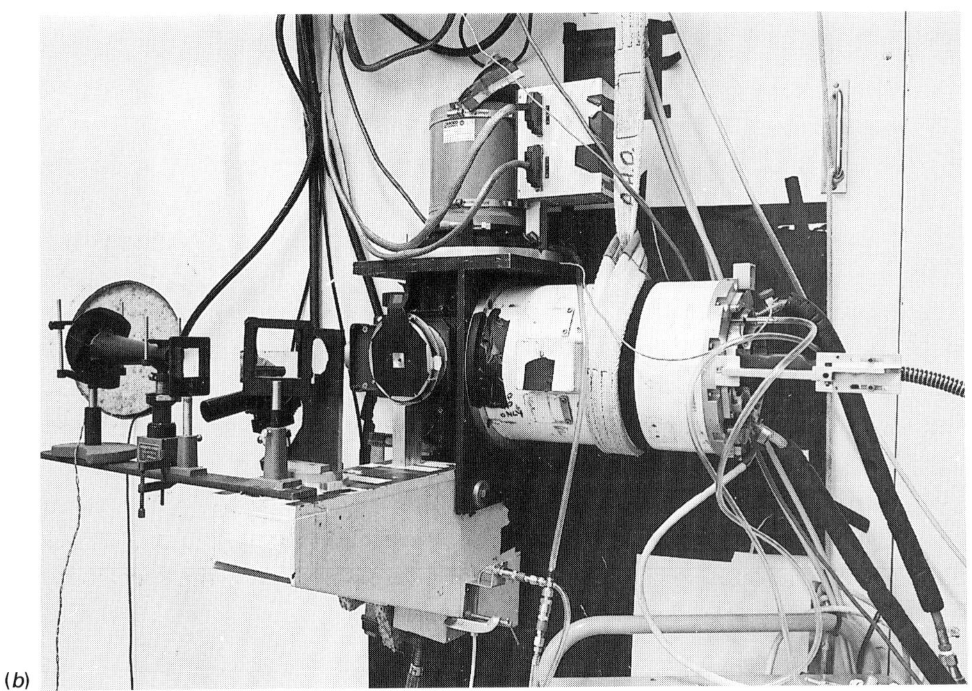

Filling the gaps

To ensure continuous coverage of the supernova, some small modifications were made to the AAT to enable it to point to the supernova throughout every night. From the latitude of Coonabarabran in New South Wales, the supernova is circumpolar – it never sets, and is visible whenever the sky is dark and clear. However, it does get quite close to the horizon – its minimum altitude is 9° and the AAT will not routinely point this low.

It is often thought that it is impossible to execute good quality observations at a low altitude like 9°. There is some justice in this view, because when we look slantwise through the atmosphere to a star at this altitude, we are looking through six times as much air as in looking at a star overhead. The tremors of this much air blur the star's image, the slanting path of the light spreads the star image into a long thin rainbow of colours, and the extra thickness of the atmosphere dims the star, especially in the blue and ultraviolet regions of the spectrum. So the AAT had been designed not to point this low – if your star was as low as the supernova sometimes appeared, you were expected to wait until it got higher in the sky, sometime later in the year.

Since the supernova was a unique, ever-changing event this philosophy was not sensible: even poor quality observations would be better than none. The engineers at the AAT altered the telescope so that it would point to the supernova whenever it was above the horizon.

ESO thought the same way – observe the supernova whenever possible! But in this case, it was ESO which improvised the solution – not with string and sticky tape, but with rubber bands.

Between May and July, only one telescope, the 61 cm telescope of the Astronomisches Institut of the Ruhr-Universität of Bochum, West Germany, was able to take spectra of the supernova when it was at an altitude of 10° (Hanuschik *et al.* 1988). The Bochum telescope had never before been pointed at this altitude (certainly not intentionally – perhaps only through a malfunction).

In order to look through the telescope's eyepiece without hanging onto the telescope structure and shaking it, one astronomer, R.W. Hanuschik, invented an out-of-the-ordinary method. He installed a rubber rope at the dome wall and, suspending himself on the rope and balancing on top of a step ladder two metres above the ground like a mountaineer, he carefully moved his eye towards the eyepiece which was at an 'impossible' position on top of the telescope tube. In this way he was able to see the supernova in the telescope, moving the telescope so that the supernova's image fell into the spectrograph and could be recorded (see Plate 6 for the results).

In another improvisation, ESO astronomers recorded their computer data on an unusual hard copy device – tracing paper. The 61 cm Bochum telescope was equipped with a very good spectral scanner which digitally recorded its data, showing it to the astronomers as an interim measure on a Visual Display Unit (VDU). There was usually not much urgency about making an immediate graph of the data – what did it matter if the data was not available that night when the star under study would be there for years? But the rapidly changing supernova spectrum made it imperative that some hard copy was available to pore over. The German astronomers stuck tracing paper over the VDU and drew the interim display with a pencil. This led to arguments on the mountain about how the supernova was changing as the hand-traced Bochum data was compared with the other astronomers' spectra and found to be contradictory. The arguments were later resolved when a machine-plotted hard copy became available from the data bank.

Information exchange

Information exchange during rapidly changing events is always a problem. Supernovae are no different from other fast-breaking news stories from remote areas, in this respect. Astronomers' most rapid routine way of disseminating information has been word-of-mouth – face to face, or over the telephone – and by writing letters, either telexes or postcards which rely on the mail. They were just beginning, by 1986, to use 'electronic mail' in which astronomical messages are transmitted through computer networks, like stock exchange information or airline reservations. Some experimental electronic mail systems had started for astronomy just before the supernova happened, but they were not in general use. The supernova sparked astronomical communications experts into prompt activity.

Nine of the postcard Circulars about the supernova were created by the IAU's Central Bureau in four days, including three on Saturday February 28. This was a record. The Director of the Bureau, Brian Marsden commented on the 'crisis situation' caused by SN 1987A that this was the first time in at least 30 years that the Circulars had been issued on Saturdays. He set up computer service accounts for new subscribers and put the Central Bureau computer onto the international Space Physics Analysis Network (SPAN).

SPAN is a packet-switched computer network organised by the US National Space Science Data Center and connecting the computers of space physicists and astronomers. A 'packet' is a group of electronic data which is sent via the network from one computer to another. A message, or even a supernova spectrum in the form of digital data, can be broken down into packets. The packets are then sent via telecommunications circuits to their destination, where they are re-assembled into the original form. Astronomers overseas could log into the Central Bureau computer and read the Circulars by electronic mail, rather than by waiting for the postman. New SPAN connections were made to the CTIO in Chile and the AAO and Mt Stromlo Observatory, in Australia. In Alice Springs and Woomera, portable Mackintosh computers were connected to the SPAN network in order to ease communication amongst groups of scientists visiting these places to launch experiments in rockets and balloons to study the supernova (Thomas, Green and McLendon 1987).

At the Space Telescope Science Institute (STScI) in Baltimore, Maryland, scientists established an information exchange – a scientifically controlled Bulletin Board – for SN 1987A. This electronic mail bulletin board is a computer file into which notices may be passed for everyone else to read, or for named individuals (a 'mail-drop'); the STScI bulletin board now contains reports from observatories and satellite missions, outlines of planned observations, reprinted IAU Circulars, a carefully maintained bibliography and a directory of astronomers as well as informal and not-to-be quoted interim results and draft papers. It is available not only through SPAN but also other commonly used networks, such as TELEMAIL, BITNET etc. (Lasker *et al.* 1988).

Archives

Another problem in studying the supernova is the continuity of the study. Most astronomy is nowadays organised in small studies aimed at answering a specific question. The answer may be a step on a longer ladder, but the study is self-contained. It might be possible to organise some supernova studies in this way, but some of the data have to be obtained long term – to follow the supernova's light curve, for example, will take frequent observations for perhaps a decade. To be useful, the data should be kept together in an archive, and not scattered in the personal filing cabinets of whoever happened to be at the telescope at the time the data were collected.

This is not a problem, of course, with an observatory with a permanent staff who can be dedicated to a given task; they can agree to work together and pool their efforts, making a permanent and continuous record. This was the approach of the SAAO which is a government-run observatory with a permanent staff with long term access to its telescopes. Exploiting the excellent sky-conditions at Sutherland where the telescopes are located, and the skills which the staff had accumulated in the science of photometry (measuring the brightness of stars – light and infrared), the observatory has built up a unique record of the total luminous output of the supernova since the outburst.

The approach of the AAO to the problem of data continuity was to set up a supernova archive of data. The archive was organised by Raylee Stathakis. The observations in the archive were mostly obtained during the one hour override which operated on the AAT from the time of the supernova explosion. The archive was set up in part to compensate the community of astronomers in Australia and the UK for the time which they had 'lost' to the supernova. Making the observations available to everybody gave them something back.

The raw data are sent from the telescope in Coonabarabran to the AAO Laboratory in Epping, where they are reduced and stored as a data file in the AAO computer. The AAO makes few photographic observations itself, but other observatories' photometric measurements are collected as they are published. An index lists all the observations, and can be queried via the international packet-switched networks such as SPAN†.

The AAO has established an archive principally for its own data. Stan Woosley of the University of Texas has called for an international archive of all the supernova data to be set up. When future theoreticians go back over the calculations they may want to test new thoughts against subtleties in the observations. The observations should not be allowed to get lost.

† Log in to AAOEPP: : [LMCSN]. From the UK STARLINK network the index is accessible as SNINFO.LIS in the AAOSERV account in the VAX computer at the Rutherford Laboratory (RLVAD).

4
The star that exploded

The LMC is one of the few external galaxies which is so near that astronomers have little trouble in identifying and studying individual stars in it. In fact, one of the reasons for the importance of the telescopes in the southern hemisphere – Australia, South Africa and Chile – is that only from these countries can the LMC be well studied. It lies near to the south celestial pole and never rises above the horizon of the northern countries.

As the nearest supernova known for centuries, SN 1987A gave the first chance that astronomers have had to identify the particular star which exploded. There had been previous extragalactic supernovae which had exploded in particular spiral arms or star clusters in other galaxies, but the actual progenitor star had never been distinguished. SN 1961V in the spiral galaxy NGC 1058 was an interesting case which nearly made it. Its peculiar appearance and behaviour could be attributed (Fesen 1985) to its detonation within a dense, bright spot on a nebula, but the star which exploded was not identified. SN 1987A was the first supernova whose progenitor was identified – within days of its discovery. Astronomers could hope that some old observations of this star could be found to shed some light on what a presupernova looked like. This hope was fulfilled – and there were some surprises.

Which star exploded?

An Australian amateur Mati Morel (as communicated to the IAU Central Bureau for Astronomical Telegrams by McNaught (1987b) was the first to notice that at the site of the supernova on old photographs was a star known as Sanduleak −69 202 (Sk−69 202). It had been listed in a catalogue

Fig. 10(*a*) The star which exploded as SN 1987A was Sk—69 202, identified here by an arrow as the progenitor of the supernova. The photograph was obtained with the 3.6 m telescope of ESO. Wisps of nebulosity cover the whole field of the plate, and clusters of stars can be picked out (several clusters are in the top left corner). The nebulosity and virtually all the stars visible in the photograph belong to another galaxy outside our Milky Way, namely the LMC, and are at the same distance, 170 000 light-years. Sk – 69 202 is one of the brightest stars on the picture – it is a supergiant star. © 1979 ESO.

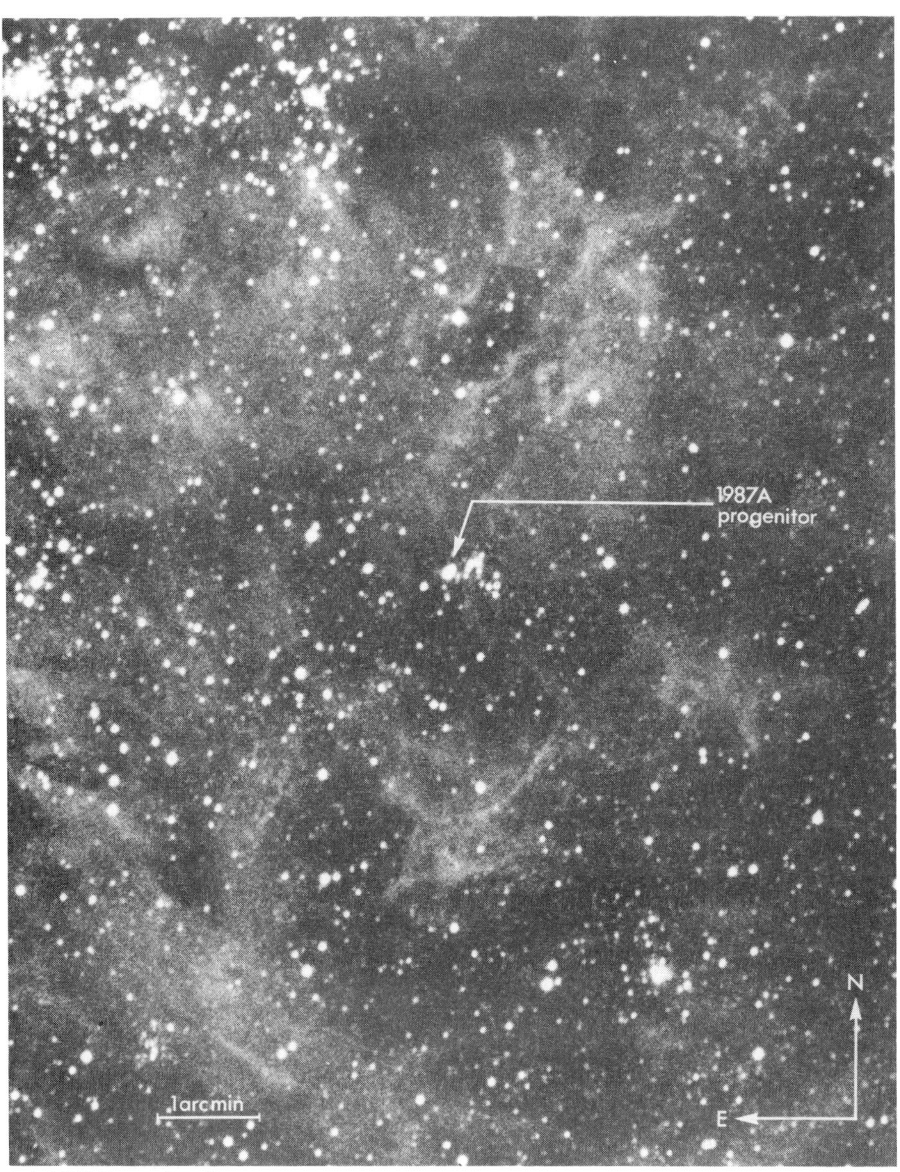

(*a*) 1979 December 6

(*b*) The same region as (*a*) is shown in a photograph taken with the ESO 1 m Schmidt telescope. The supernova had reached visual magnitude 4.4; the cross over the supernova is an artefact of the telescope caused by diffraction around the cross-shaped supports of the plateholder. © 1987 ESO.

(*b*) 1987 February 26 01 : 25 UT

compiled by Nicholas Sanduleak (pronounced Sand-yóu-lee-ak) of 1272 hot, blue stars, which were members of the LMC. Shara and McLean (1987) and Sanduleak (1987) himself followed up McNaught's communication and suspicion focused on this star as the progenitor of SN 1987A.

Sk – 69 202 was a bright, blue star – a hot supergiant. The different types of stars are coded by astronomers as letters, numbers and Roman numerals which represent the appearance of their spectra. This star's observed spectral type was B3I – other circumstantial evidence, such as its brightness, would place it on the bright side of the supergiant range and it could be thought of as class B3Ia. This code for the appearance of the spectrum of the star translates to the equivalent of saying that the surface temperature of the star was 28 000 K and its optical luminosity about 40 000 solar luminosities.

Astronomers at several observatories immediately tested the suspicion that the supernova and Sk – 69 202 were the one and the same star. The test was by astrometry – accurate measurements of the positions of the supernova and of Sk – 69 202.

A typical one of these tests was made at the Anglo–Australian Observatory. White and Malin (1987a, b) found many AAT photographs of the region of the supernova, taken before it had exploded, and selected two of the sharpest (good 'seeing') from 1983 and 1984 for special study. Two further photographs, showing the supernova, were made on 1987 February 27 (one of them only a 15 s exposure because the supernova was so bright!).

The positions on each photograph of the images of the supernova and of Sk – 69 202, and other stars near to the supernova site, were determined with a machine which measures positions to an accuracy of 1 micron (the thickness of a hair). In addition, several stars ('secondary standard stars') in the region up to 0.5° from the supernova were measured on these four photographs in order to relate the supernova and Sk – 69 202 to a common framework of star positions. The positions of the secondary standard stars were themselves determined by measuring wider field photographs (taken in the mid 1970s) which contained the images of brighter stars whose positions had been measured in 1969 by astrometrists and tabulated in the so-called Perth catalogue ('primary standard stars'). It was necessary at each stage to account for the motions of the primary standard stars in the intervals between the photographs – the errors in the quoted measurements of the motions of the primary reference stars were the cause of the biggest uncertainties.

The supernova itself was so bright that, rather than measure the image of the star itself, the two Australian astronomers measured the diffraction spikes on the image. These four spikes which radiate north–south and east–west from

Fig. 11. David Malin of the AAO has superimposed a negative image of an old photograph taken with the 3.9 m AAT on a positive image of the supernova. At the intersection of the diffraction cross lies the centre of the white image of the supernova. It coincides with the black image of Sk − 69 202. © AAT 1987.

bright star images are artefacts of the telescope: the plateholder and, indeed, the camera operator, in this case David Malin himself, are installed within the telescope beam in a cylindrical container known as the 'prime focus cage'. The cage is held to the telescope tube (actually an open steel-work structure) by four struts which point north–south and east–west. The shadow of the struts removes some light from the beam from the star. Diffraction of light within this particular shaped aperture creates an image of the stars which is attractively 'star-shaped', like the conventional image of the Star of Bethlehem on a Christmas card†.

Astronomers usually regard these diffraction spikes as a nuisance, since they destroy the circular symmetry of the star images and make them difficult to analyse, but in this case the spikes on the supernova image were a boon. They were fainter and finer than the supernova image and enabled its position to be measured more accurately.

Several other observatories measured the positions of the supernova and the stars near Sk − 69 202 (see summary by West (1987)). There were three stars within 4″ of the supernova – they were labelled stars 1, 2 and 3. Star 1 was Sk − 69 202 – it was the brightest. As noted by Sanduleak (1987) and Lasker (1987) Sk − 69 202 was double, with Star 2 three magnitudes fainter and 3″ away to the north-west – it shows as an elongation of the image of Sk − 69 202. A third component to the image (Star 3) was detected on plates taken at CTIO (Chu 1987, Walborn *et al.* 1987) and is 1.4″ south-east of Sk − 69 202 and not much fainter than Star 2 (Fig. 12).

Figure 13 was prepared by the ESO and shows a computer representation of a photograph of Sk − 69 202. The image of the central star (Star 1) was fitted by a mathematical function which represented the shape of a single star – it could not take account of the presence of the other stars nearby. Star 1 was then subtracted from the computer image: the traces which remain are inaccuracies in the fitting of the mathematical function to the data, including the images of Stars 2 and 3. It is virtually certain, according to van den Bergh (1987), that all three stars are related and within the same association, although they are separated by a light-year or more.

All observatories showed that the supernova and Star 1 but not Stars 2 or 3 were at the same position in the sky; the best measurements colocated them to within about 0.05″ in Right Ascension and Declination. The same accuracy in latitude and longitude on the Earth would identify a particular room in a house

† A TV cameraman may sometimes, for effect, deliberately install a cross-shaped obstruction over his camera lens in order to produce a similar 'highlight' as his camera views studio lighting.

(25 square metres of the Earth's surface): there would be a strong implication that two people with the same latitude and longitude to this accuracy were actually just one individual, or, at the very least, closely related.

Thus there was a strong presumption that the supernova and Sk – 69 202 were the same star, or at least members of the same star system. The

Fig. 12. Sk – 69 202 is the brightest star in this pre-supernova, highly magnified photograph from ESO. The star images are not exactly circular due to less than optimum equipment performance. Although all the star images are elongated to the north-east (upper left corner), Sk – 69 202 is elongated to the north-west by the image of a companion star (it has a profile more like a cottage loaf than a rugby football). The companion is Star 2. Moreover, there is another companion, Star 3, less readily identifiable by the extra fuzzy eastern (left) edge to the image of Sk – 69 202. © ESO.

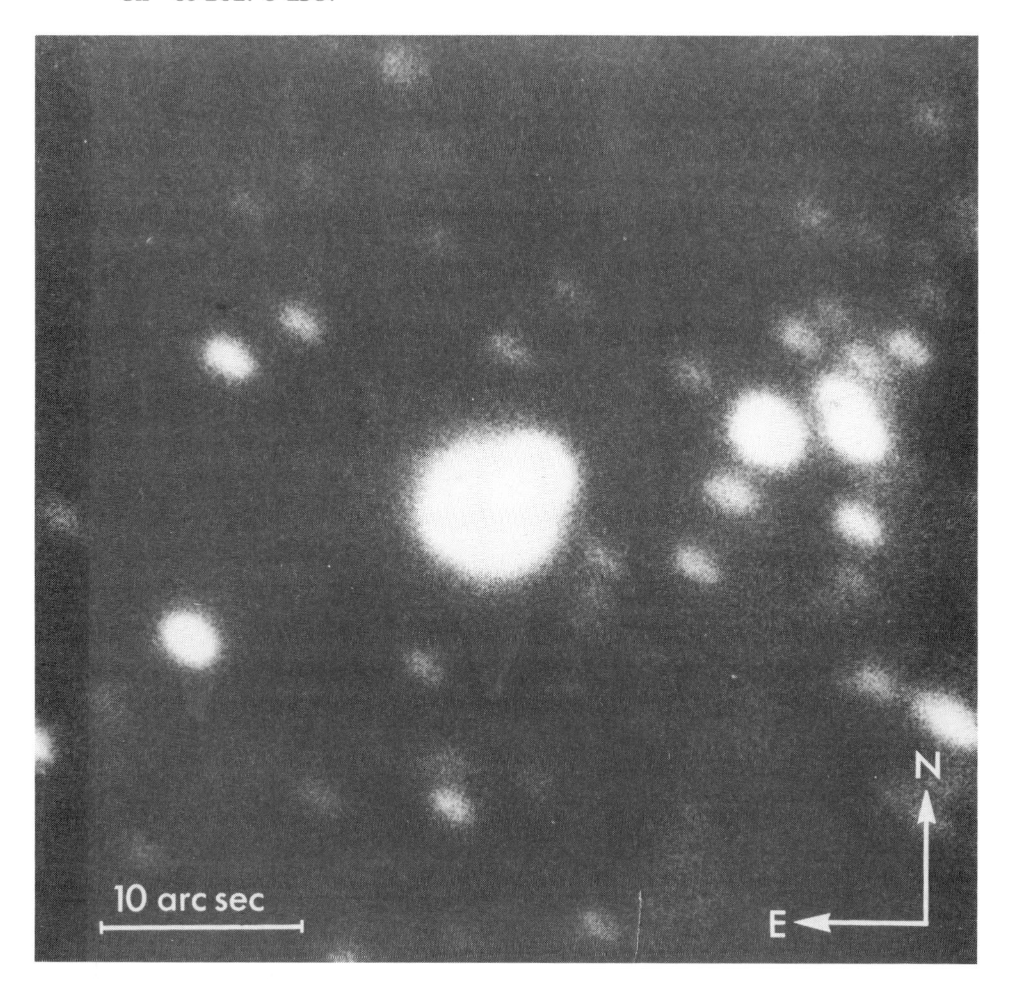

coincidence of position could not be a complete proof of this, however. For example it could be that many people living above one another in a block of flats would have the same latitude and longitude. Similarly, it could be possible, but unlikely (as noted by van den Bergh (1987)), for two stars in the LMC to lie at the same position in the sky, one thousands of light years in front of the other, without any connection between them.

The simplest conclusion from the astrometry, however, was that Sk – 69 202 had been replaced by the supernova: Sk – 69 202 was dead.

'Sk – 69 202 has not exploded'

In a sensational development, however, this conclusion was overturned – temporarily – by results from the IUE astronomy satellite by both

Fig. 13. The photographic image of Sk – 69 202 in Fig. 12 was read into a computer to examine the star system in detail. The brightest star, Star 1, was removed from the computer image, represented here as a contour diagram from ESO. The removal took into account the elongation of the star images in the rest of the photograph. The images of Stars 2 and 3 clearly remain, together with traces of Star 1 which have not completely been subtracted away.

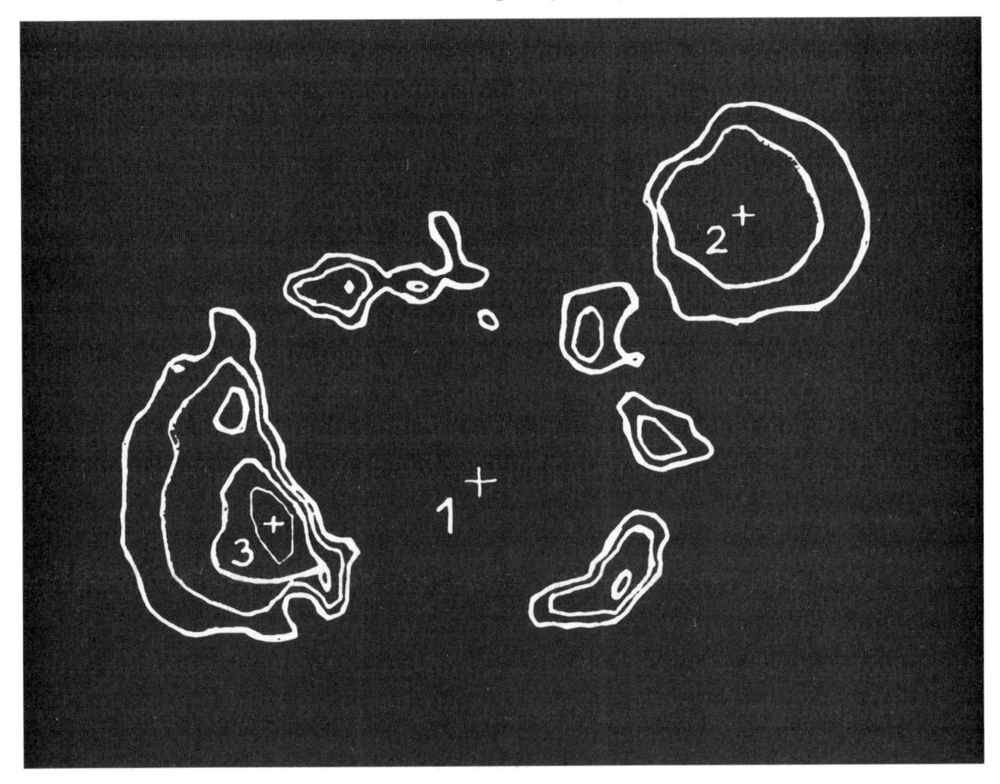

the European IUE supernova team (Gry, Cassatella, Wamsteker, Sanz and Panagia 1987; Cassatella, Wamsteker, Sanz and Gry 1987) and, in the same week, by their American colleagues and rivals (Sonneborn and Kirshner 1987a).

The ultraviolet light from the supernova faded away rather rapidly (much faster than previous supernovae had faded) – at the rate of 0.1 magnitude per hour from February 25–7 at 1325 Å, near the wavelengths to which IUE was sensitive. A loss of 0.1 mag represents a fall in brightness of 10%, which multiplies over 72 hours to a fall to nearly 1/1000 of the brightness when first detected. But as the supernova became invisible to IUE at 1325 Å there remained a faint residual ultraviolet light – after five days (March 2) the very rapid decrease in ultraviolet light had stopped completely. The spectrum of the residual ultraviolet light resembled an ordinary star of the kind that Sk – 69 202 had been. Apparently that star had not exploded after all, since its light could still be detected. Moreover, there were signs that there were two stars contributing to the residual spectrum; since Chu (1987), who discovered Star 3, had estimated its brightness at 17.5, it did not seem plausible for it to be detected by IUE and the two stars were thought to be Stars 1 and 2.

On March 4, the European team were convinced that both Sk – 69 202 and the supernova could be seen at the same time, with the supernova's light dominating at wavelengths longer than 2500 Å. A detailed comparison with spectra from the IUE archive in the range 1200–1650 Å, they reported, 'confirms the similarity of the recent spectra with an early-type supergiant in the LMC.' The US team reported their identical conclusion on March 7. 'IUE spectra taken on February 27, 28, March 1 and 2 confirm the detection of Sanduleak – 69 202 beneath the fading far-ultraviolet flux of SN 1987A,' they wrote and concluded that Sk – 69 202 had not exploded.

Some astronomers were particularly ready to believe this because of a prediction from the theory of supernovae. It had become the accepted theory that supernovae occurred, not in blue supergiants, but in red supergiants. The reason for this was that theoreticians had difficulty in making supernovae as bright as they usually were unless the explosion took place in a very large (and therefore red) star. Very early, McNaught had come up against reluctance to believe Morel's discovery, which McNaught had communicated, that the blue supergiant Sk – 69 202 was the progenitor of SN 1987A, and the IUE results intensified the scepticism.

It would be a cruel joke by nature to place SN 1987A so close in the sky to Sk – 69 202 totally by chance – in fact White and Malin (1987a, b) had estimated the probability that this would occur as less than 1 chance in 10 000.

Astronomers still worked on the hypothesis that there was some relation between Sk – 69 202 and the supernova's progenitor, but they began to contemplate the possibility that a red companion star in Sk – 69 202's star system was the progenitor of SN 1987A.

Another star in the Sk – 69 202 system?

In their desire to find a red supergiant, observers began to examine the records of the red light from Sk – 69 202 to see if there was one hidden in its blue glare. Humphreys, Jones, Davidson, Ghigo and Zumach (1987) of the University of Minnesota had given a preliminary report on March 1 that colours for Sk – 69 202 measured off old plates were slightly too red for a B-type supergiant. The implication was that a red star was present to emit this red light.

Blanco (1987a) of CTIO had followed this up by March 23 with more accurate measurements. He calibrated five old red and four old near-infrared presupernova photographs showing Sk – 69 202 with new photoelectric measurements of stars in the vicinity of the supernova (there was no reason for other stars in the vicinity to have changed in the last 20 years and their brightnesses observed now could justifiably calibrate old photographs). He concluded that it was just possible that Sk – 69 202 could have a close reddish companion of the kind which could explode as a supernova; it would have to be a rather faint supergiant, and not very red (K2Ib was the code for its likely spectrum at the brightest that could be accommodated). When Blanco (1987b) reconsidered this and other evidence later he maintained that there was a slim possibility that there was such a star in the Sk – 69 202 system but he concluded that this was stretching the available data to the limit.

West, Lauberts, Jørgensen and Schuster (1987) tested the possibility that Sk – 69 202 was not the progenitor by looking for eclipses of that star by an unseen dark companion. If there was one, it might have been the supernova progenitor. No eclipses were visible as a dimming on Sk – 69 202 on any of 40 photographs taken at ESO. This confirmed Hazen's (1987) statement based upon a comprehensive collection of photographic plates taken to search for variable phenomena that 'No major change in the brightness of Sk – 69 202 was found in a search of the star's area on 502 Harvard blue patrol plates taken 1899–1953' or on 59 plates taken between 1971 and 1986.

Testor and Lortet (1987) of the Observatoire de Paris, Meudon, looked out from their collection some six year old electronographic pictures and thought that the pictures showed Sk – 69 202 'was composed of at least two stars aligned north–south (within 1.0″) and might even be a cluster of stars'. There seemed to be room for many other stars around Sk – 69 202 in this observation, but it was never confirmed.

The survivors

The question of which star had exploded was settled by definitively identifying the survivors of the supernova explosion. The star that was missing would clearly have been the progenitor. Because the supernova fireball was so bright – thousands of times brighter than Sk – 69 202 – it would be some time before the survivors become visible optically. But the supernova faded very rapidly in the ultraviolet and the likely survivors would shine brightly in the ultraviolet: thus the IUE was the best instrument with which to identify the

Fig. 14. Spectra of the ultraviolet light (1300–1650 Å) of SN 1987A at the time of the outburst, and some three weeks afterwards, were obtained by IUE. The range of wavelengths is spread in the long direction of the spectra. Perpendicular to this, across the thickness of the spectra, is represented the width of the images of the stars whose spectra are being recorded. The spectrum of the supernova itself is thin – only one star is responsible for this spectrum, whose exposure was too brief to record any other fainter ones. But the longer exposure spectrum taken three weeks later is wider, and is even separated clearly in some regions into the spectra of two stars. These are Stars 2 and 3, remaining after the explosion. Star 1, Sk–69 202, has disappeared. Courtesy of W. Wamsteker, IUE tracking station, Vilspa, Spain.

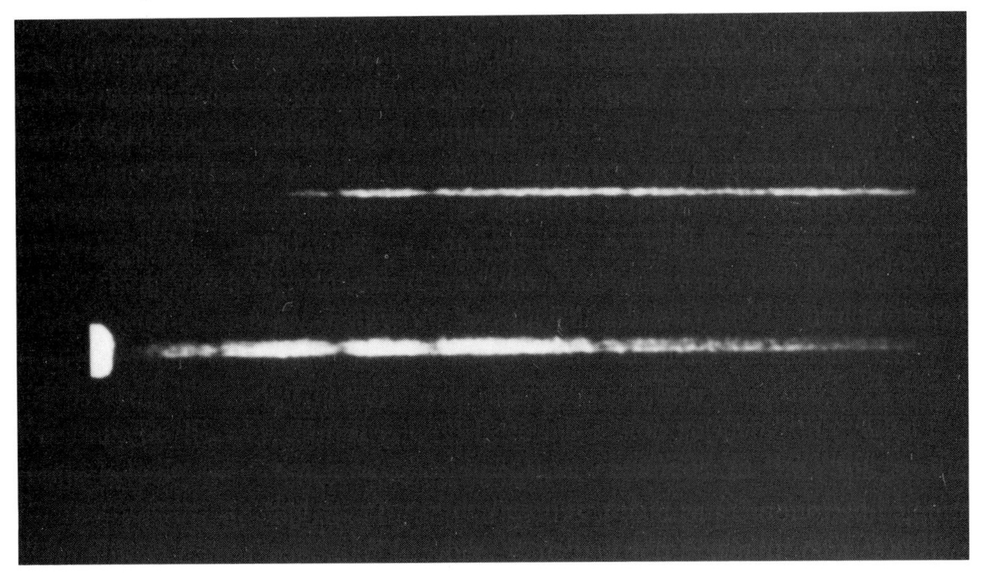

Fig. 15(*a*) A cut across an IUE spectrum like the later one in Figure 14 can be fitted very well (the open circles lie on the solid line) with the sum of two star-profiles (the dotted lines) representing Stars 2 and 3.
(*b*) A cut across a spectrum, taken at a time when SN 1987A was of comparable brightness to Stars 2 and 3, cannot be well fitted by the sum of two star-profiles (the open circles lie above and below the solid line) . . .
(*c*) . . . but the same data (open circles) as in (*b*) can be fitted by three star-profiles, the middle one representing the supernova and the ones on the left and right Stars 2 and 3 respectively. The profiles in (*a*) and (*c*) which represent Stars 2 and 3 are constant in separation and brightness, but in (*a*) the supernova has faded out of sight. Data and analysis by George Sonneborn, Bruce Altner and Bob Kirshner.

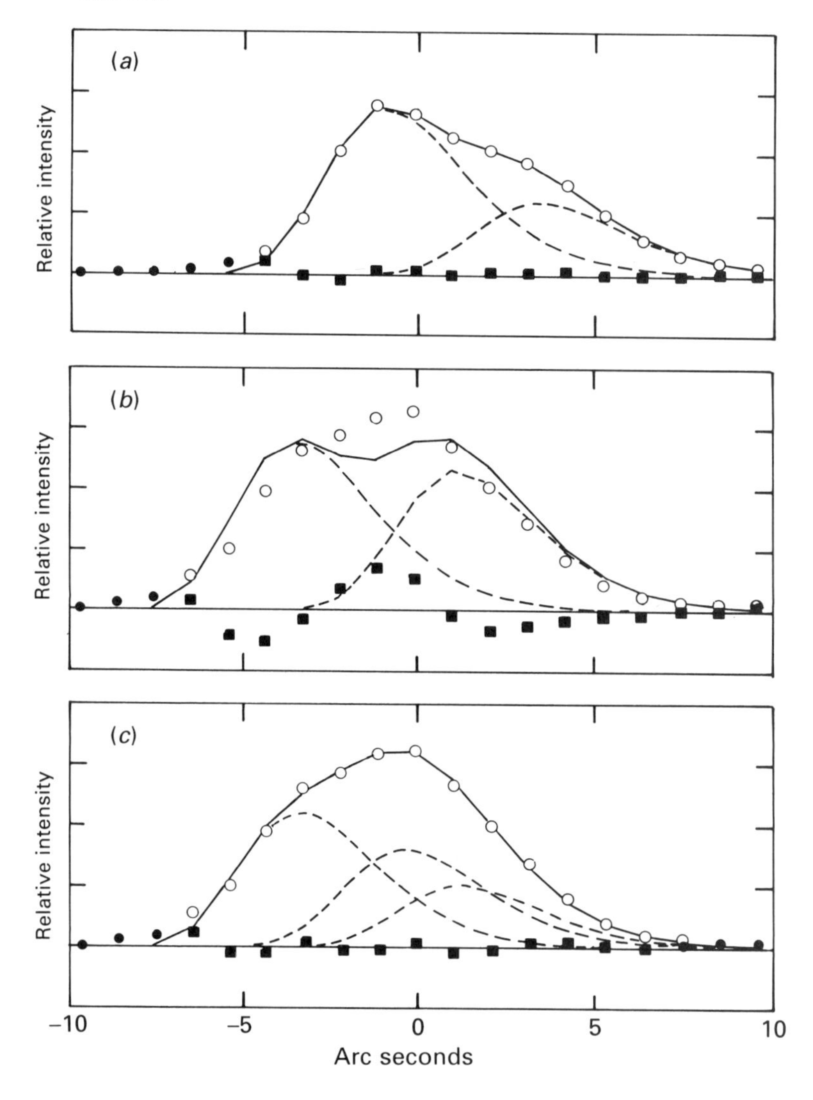

survivors. Ironically, IUE was the instrument which had cast doubt on the identification of the supernova with Sk − 69 202 in the first place.

Gilmozzi (1987) reporting on behalf of the European IUE Target-of-Opportunity Team for Supernovae (Gilmozzi *et al.* 1987) and Sonnenborn, Altner and Kirshner (1987) gave a careful discussion of this point.

Once the mapping of the three stars (Sk − 69 202 itself and Stars 2 and 3) within 4″ of the supernova had been completed, it was possible to examine how the three might have contributed to the spectra seen by IUE after the supernova had faded. The crucial point was that the spectra of stars obtained with IUE are taken through a large entrance aperture (10″ × 20″) so that all three stars could contribute to the spectrum which was recorded. There were, in fact, only two stars whose spectra could be seen, and their separation was 4.5 ± 0.2″. This could not fit with Stars 2 and 1 (separation 3″) but could well be Stars 2 and 3 (separation 4.4″). In addition, the two spectra showed that the brightness of the two stars in the region 1300–1500 Å was in the ratio 1.75 ± 0.12 : 1. This could not fit with Stars 2 and 1 (ratio 0.06), but could be Stars 2 and 3 (ratio 1.4).

Roberto Gilmozzi, on behalf of the European team, concluded that, barring exotic interpretations ('Star 1 is behind the SN' or 'Star 1 is inside the SN'), Sk − 69 202 had disappeared because it was the actual progenitor of SN 1987A. Kirshner, on behalf of the American team, concurred. The report that Sk − 69 202 was still there was a false alarm. Any residual doubt was dispelled by March 1989 when the supernova had faded and Stars 2 and 3 were seen in conventional optical telescopes, but not Star 1.

Kirshner, with his ebullient self-awareness, and his sense of humour and love of science, found (Bates 1988) a one-liner to sum up the confusion: 'The supernova was a bit dim to start with – like some of the observers.'

Description of the progenitor of a supernova

There are literally millions of stars of the brightness of Sk − 69 202 or brighter. It is a tribute to the patience of astronomers that anything at all is known of the progenitor of SN 1987A, an unexceptional star before the outburst of 1987 February 23 called such dramatic attention to it. The outburst sent astronomers scurrying to their records for any significant facts.

The story gradually emerged (Blanco 1987b) – parts of it have already been given in the first chapters of this book.

Photographs of the LMC which had been taken through a thin glass wedge (an 'objective prism') showed the individual spectra of thousands of stars. Sk − 69 202 was type B3I – a blue supergiant star. Its magnitude fitted it to be a rather bright supergiant; the code for this is B3Ia.

It was not significantly variable – there were no signs of eclipses (no intervening companion star), nor brightening nor irregular fluctuations (no instability developing before the supernova explosion).

It was not an especially bright infrared source, neither as observed from the ground at wavelengths of 1 micron or so (no second, red star companion) nor from the Infra Red Astronomy Satellite (IRAS) in space (no signs of ejection of warm carbon grains as the star passed through an advanced state of evolution).

There was, in fact, little evidence before it exploded which would have enabled astronomers to pick it out as a soon-to-explode supernova.

5
The life of a supernova progenitor

The discovery of supernovae

There are two broad kinds of supernovae, unimaginatively called Type I and Type II. They are recognised by their optical spectra. From this operational definition has followed a painstaking process by which the origins and properties of each type have been discovered.

At first, up to 1933, supernovae were discovered only by haphazard inspection of the sky. Someone would point a telescope to a galaxy, perhaps to take a photograph of it, or study some feature in it. In comparing the current image of the galaxy with an older photograph, either through the telescope itself or in the new photograph, they would notice that it contained a star which was not present before. Such a new star was possibly a supernova. (Other explanations for this phenomenon would have to be ruled out, such as the possibility that the 'star' was an asteroid or minor planet which had intruded by chance onto this part of the sky, or that the apparent supernova was a variable star which was ordinarily below the threshold of detection but which, after a relatively small increase in brightness, had become more prominent.)

Supernovae are still discovered like this. But most of astronomers' knowledge has come from studying examples of supernovae found as a result of systematic searches.

The first such searches were organised and run by Fritz Zwicky, starting in 1934. Repeatedly, he and his successors photographed fields of nearby galaxies and looked on the photographs for supernovae. Zwicky began his searches with relatively small astronomical cameras, but inspired the construction of the world's largest camera to further his discoveries. It was the 48 in Palomar

Observatory Schmidt telescope which was used from 1958 to 1975 for three moonless nights each month to photograph 38 fields containing 3003 galaxies in clusters and groups. Out of the total of 281 supernovae discovered up to then by the Palomar supernova survey, 178 were discovered with this camera.

The supernova searches gave an idea of the rate at which supernovae are discovered in an average galaxy. For example, if, as above, 178 supernovae were discovered in 17 years in 3003 galaxies, then the observed rate of supernovae of both types in an average galaxy is 1 every $17 \times 3003/178$ years – i.e. every 280 years. But this is not a very useful number, because there are many systematic effects to account for in correcting for the supernovae which are missed. For example fewer supernovae are found in the centres of galaxies than at the outer edges, even though there are many stars in the centres of galaxies. This is because supernovae are harder to spot amongst the bright mass of stars. Correcting for systematic effects like this, Gustav Tamman deduced that supernovae occur at the rate of 1 every 20 years in a galaxy like our own.

Astronomers have found this rate hard to reconcile with the fact that we have not seen a supernova in our Galaxy for over 300 years, nor a supernova in the Andromeda Galaxy (our Galaxy's nearby twin) for 100 years. Where have all the supernovae gone? Why have we missed so many?

Recently, Sidney van den Bergh and Robert McClure used remarkable statistics from an Australian amateur's visual discovery of supernovae to give a better impression of the supernova rate (van den Bergh, McClure and Evans 1988).

The Rev. Robert Evans, who is a Uniting Church minister (*Time* 1987; Evans 1988), has set out to make a deliberate search for supernovae by eye.

Selecting galaxies which are close enough to give a supernova which he could see by eye in his amateur-sized 10 in telescope (magnitude brighter than 14.5), Evans has remembered a list of a large number (1017) of galaxies, their positions and their appearance in his telescopes. Every available clear, dark night he works his way through the list checking for new stars in the fields of the galaxies.

Evans recognises galaxies like the faces of his friends and he notices if there is a new star, just as he would notice a pimple on a familiar face. If he sees a new star which could be a supernova in one of his galaxies, he carries on observing the rest of the list for a half hour before returning to the galaxy. Sometimes the 'star' has moved: it must have been a minor planet and not a supernova after all. But if the new star is in the same position as before, he can check against notes and phone professional astronomers with the news of a probable supernova.

Table 3. *Amateur discoveries of supernovae (prior to March 1987)*

SN	Galaxy NGC (or other name)	Method	SN mag.	Type	Discoverer
1968L	5236 (M83)	visual	11.9	II	Bennett
1979C	4321 (M100)	visual	12.1	II	Johnson
1981A	1532	visual	13.5	II	Evans
1981D	1316 (Fornax A)	visual	12.5	Ia	Evans
1983G	4753	photograph & visual	12.8	Ia	Okazaki & Evans
1983N	5236 (M83)	visual	11.5	Ib	Evans
1983S	1448	visual	14.5	II	Evans
1983V	1365	visual	13.8	Ib	Evans
1984E	3169	visual & photograph	15.0	II	Evans, Metlova[a] & Okazaki
1984J	1559	visual	13.2	II	Evans
1984L	991	visual	13.8	Ib	Evans
1984N	7184	visual	14.0	Ia	Evans
1985B	4045	photograph	13.0	II	Horiguchi
1985G	4451	photograph	14.5	II	Horiguchi
1985P	1433	visual	13.5	II	Evans
1986A	3367	visual	14.0	Ia	Evans
1986G	5218 (Cen A)	visual	11.4	Ia	Evans
1986L	1559	visual	13.3	II	Evans
1987A	— (LMC)	photograph & visual	2.9	II	Shelton[a], Duhalde[a] & Jones
1987B	5850	visual	14.4	II	Evans

[a]Shelton and Metlova are not amateurs; Duhalde is a night assistant.

In five years' work, Evans made 50 403 observations. (It is illuminating to calculate that at one minute to complete an observation this means that Evans looked at galaxies for the equivalent of 100 working days in five years.) Evans discovered eleven supernovae in this way – and he would have discovered four more except that they were discovered first by others. The score has increased to 15 (Table 3) since 1985 when Evans was given a 16 in telescope by the CSIRO. Now he can see fainter supernovae.

Because Evans works to a well-defined sample of galaxies and works by eye his results are not so biassed as chance discoveries or photographic surveys.

For example, photographic surveys concentrate on clusters of galaxies, because then more images of galaxies are included in a single shot. This discriminates in favour of elliptical galaxies (of which there are more in clusters) and against spiral galaxies (which tend to be more isolated).

On the other hand, those supernovae which are found in spiral galaxies tend

Table 4. *Supernova rates from Evans' visual survey. Supernovae per century per galaxy of $10^{10}L_{\odot}$ (suns).*

Type Ia supernovae	0.3 per century
Type Ib supernovae	0.4
Type II supernovae	1.1
All types	2 per century

to be found in face-on spirals rather than the spirals inclined edge-on to the line of sight. This may be either because there is a greater obscuring effect by dust which hides supernovae in these galaxies, or because astronomers take more pictures of face-on galaxies (they are prettier, is one reason, and individual objects can be studied in their spiral arms, is another).

Or again, photography finds it hard to detect supernovae in the bright, overexposed central regions of galaxies, but the eye is better adapted to variations in brightness of different regions.

Van den Bergh *et al.* (1988) express the supernova rate derived from Evans' observations in terms of a standard galaxy whose light is equal to 10^{10} suns ($L_{\odot}$). They thus correct for the expectation that a galaxy which is brighter and contains more stars will have more supernovae (and vice versa). And they assume that the Universe is of such a size that Hubble's constant is 100 km/s per Mpc – it is necessary to use this to calculate the galaxies' luminosity.

They derive the supernova rates of Table 4. (See below for the explanation of the different types of supernovae.) If Hubble's constant is actually 50, in units of km/s per Mpc, the true rates are $\frac{1}{4}$ of the rates in the table. (The compromise value of Hubble's constant is half way between these extremes.)

Even though Evans' observations are a heroic achievement for one man using his spare time, the number of supernovae in the analysis by van den Bergh and his co-workers is small. The rates are subject to a large statistical uncertainty which in total could be as large as 50%.

The rates sum to an average of one supernova per 60 years for a galaxy like M31, the Andromeda Galaxy, rather than the rate of one per 21 years.

In a faint galaxy like the LMC ($\frac{1}{12}$ the brightness of our Galaxy) the Evans' statistic predicts one supernova every 625 years.

Following up supernova discoveries

Except in the rare case when a supernova is discovered before its maximum brightness, it is fading from the moment of discovery. Thus it is necessary to react quickly to its discovery, before it becomes too faint to investigate. Right from the start, Zwicky organised follow-up studies of the supernovae which he discovered. Walter Baade would measure the light curves and Rudolph Minkowski and Milton Humason would obtain spectra. They used the world's then most powerful telescopes, the 60 in and 100 in Mt Wilson telescopes.

From studies of the spectra of the first supernovae discovered in Zwicky's supernova search, Minkowski found in 1940 that most supernovae discovered (three quarters) were very much alike and he called them Type I. Since that time, astronomers have subdivided these into Type Ia and Type Ib on the basis of a detail in the spectra. The remainder were different from Type I but not all identical. Minkowski called them Type II. Some astronomers, such as Zwicky, were able to recognise further types (III, IV and V), but not all astronomers are convinced that these are distinct kinds. Type Is and Type IIs have different light curves.

Type I supernovae

A supernova of Type I has no hydrogen lines in its spectrum. It does have spectral lines of silicon (the feature at 6150Å is now the definitive feature for Type I), calcium, oxygen, sulphur, magnesium, . . . but no hydrogen lines. The implication is that the supernova's outer layers do not contain significant amounts of hydrogen, in spite of the fact that hydrogen is by far the most common element in the Universe and that hydrogen lines are almost without exception found in the spectra of all stars. Type Is must therefore be the explosion of an uncommon kind of hydrogen-free star.

There is a subdivision of Type I, namely Type Ia and Ib, based on the strength in the optical spectrum of the silicon absorption line at 6150 Å. If the silicon absorption is strong the supernova is Type Ia, if weak it is Type Ib. There are probably two kinds of hydrogen-free star which we must identify as the seat of supernova explosions. Since Type Is are all so much alike, in spite of the small detail in the spectra, we are probably looking for very similar stars for Types Ia and Ib, and the stars must be very well defined in their essential properties – practically identical in mass, size, composition etc.

Astronomers are fairly sure that they have identified the seat of the Type Ia supernovae, but are still puzzled about Type Ib s. It is believed that Type Ia s represent the total disintegration of a white dwarf star in a binary system. White dwarfs are stars which have the mass of the Sun but which are about the size of the Earth. A carbon-oxygen white dwarf star, containing no hydrogen, may accrete material from a companion star to such an extent that its central temperature suddenly rises above a critical value at which carbon ignites in an uncontrolled thermonuclear fusion runaway reaction. This theory accounts nicely for the similarity of Type Ia supernovae – all such white dwarfs would have the same mass and composition, so their explosions would be identical. The closely similar Type Ib supernovae must be explosions in stars which mimic white dwarfs in most ways, except for some details.

Type II supernovae

In contrast to Type Is which have no hydrogen apparent in their spectra, Type IIs, on the other hand, do show hydrogen and are evidently the explosions of an ordinary kind of star. Since they are a much broader class of objects, there must be considerable differences in the structure of whatever kind of star is responsible. The main clue about this point comes from a significant fact which gradually emerged from the supernova surveys: different kinds of galaxies host Type Is and Type IIs.

Not all galaxies are alike. They differ in their shape.

Some galaxies are irregular conglomerations with no particular structure. They are called irregular galaxies.

Some galaxies are featureless balls of stars – circular-looking galaxies (which might be truly spherical) or elliptical-looking galaxies (which might be elongated, cigar-like volumes, or frisbee-shaped discs seen obliquely in perspective). These are all called ellipsoidal galaxies.

The most photogenic and varied galaxies are the spiral galaxies. They are flattened discs. When seen face-on they show attractive spiral arms. The arms can be many, tightly wound and almost circular, or there can be just two arms showing a pronounced spiral shape. The arms may start close to the nucleus of the galaxy, or from the ends of a bar of stars lying across the nucleus.

The nucleus of a spiral galaxy may be a relatively minor part of the galaxy, or the galaxy may have a large nucleus and minor spiral arms. When seen sideways on, spiral galaxies show more clearly the relative proportions of the central nucleus and the spiral arms in the flatter disc which surrounds the nucleus.

Type I supernovae are found everywhere in all kinds of galaxies, while Type IIs occur only in the arms of spiral galaxies.

The significance of this point is that the spiral arms of a galaxy are the locations where hot, bright stars are found. Hot, bright stars occur in clusters; because they are hot, they emit ultraviolet light which ionises interstellar gas to make nebulae. The clusters of bright stars and the nebulae delineate the spiral arms.

The LMC is a spiral galaxy – poorly defined, because it contains so few stars, but full of hot, bright stars and nebulae – and the LMC supernova occurred near the brightest nebula in the galaxy. When the supernova's spectrum was first recorded, it had clear (though at first misidentified) hydrogen lines in its spectrum. Thus SN 1987A is a Type II supernova. It is consistent that it occurred in the LMC.

To show the significance of this I have quickly to review how stars live their lives.

Star formation

Stars form from gas clouds (principally hydrogen) in the space in a galaxy. The gas clouds become unstable and fragment into numerous smaller bodies – protostars.

As in virtually any process in which a big thing gets broken up, smaller fragments are more common than the larger ones, so small stars are more numerous than large ones. The fragments contract under their own force of gravity and form spherical globes. The compression of the protostars heats the globes of gas, just as compression of air in a bicycle pump heats the tyre. The surfaces of the protostars warm and begin to radiate infrared. The hydrogen atoms of which the stars are principally made break up as they collide forcefully in the temperatures which are generated inside the stars – the hydrogen atoms become free electrons and protons.

If the protostars are big enough, the central regions of the globes heat further, to a hot enough temperature that the free atomic nuclei, such as protons, collide with each other. This starts nuclear reactions. The nuclear reactions release further energy which flows to the surface of the star. Here it is radiated as light and is the means by which we perceive the star.

Protostars which are smaller than, say, $\frac{1}{20}$ the mass of the Sun never reach this temperature: they never release energy generated by nuclear reactions, only the limited energy from gravitational contraction. They quickly fade from the scene. On the other hand, protostars which are bigger than say 100 times the mass of the Sun heat up too much and disintegrate. Stars thus lie in mass between 100 times and $\frac{1}{20}$ the size of the Sun.

The pressure balance in a star

When the nuclear reactions begin, the star stabilises – the contraction due to gravity ceases. This is because the star becomes a stable balance between the inward force of gravity and an outward pressure force. The pressure comes from the energy which is generated by the nuclear reactions. The energy, which these reactions release, heats the material within the star and the collisional force of electrons and protons within this material constitutes one such pressure force, called particle pressure. The other force is radiation pressure which is caused by the push of radiation as it jostles outwards and upwards in the material of the star.

The physical laws which connect the pressure, temperature and density of the material in the star are called equations of state†. The radiation pressure and the particle pressure, governed by the appropriate equation of state, constitute the outward forces which keep the star in equilibrium against the inward force of gravity.

If too much energy is created by the nuclear reactions, the pressure rises and the star expands. The star cools and turns the nuclear reactions down. On the contrary, if too little energy is created, the star contracts some more, heats itself up and turns the nuclear reactions up. The balance is self-regulating. The star is said to be in hydrostatic equilibrium.

The colour-magnitude diagram

These relationships have the consequence that stars are not arbitrary in size, surface temperature, energy released, mass etc: there are definite connections amongst the parameters of a stable star. This shows when astronomers plot one parameter against another for collections of stars.

One of the most useful plots is to represent stars as points in a diagram of luminosity versus surface temperature††. This goes under the names of

† Boyle's ideal gas law $P = \rho RT$, which connects the pressure, P, and temperature, T, of gas with its density, ρ, under the conditions normally found on the Earth's surface, is a simple example of an equation of state. The constant R is called the ideal gas constant, determined experimentally or from other more fundamental physical laws.

†† This combination of parameters into a diagram is useful because luminosity is related to the brightness or magnitude of stars, and surface temperature is related to the spectrum or colour of stars – with a certain amount of hard work, astronomers can measure these quantities, and they are observables. It might be theoretically more revealing to plot central density versus diameter (to pick an arbitrary combination of stellar parameters), but this would not be practical.

Hertzsprung–Russell (H–R) diagram, after the names of the two astronomers who invented it, or the colour–magnitude diagram for the reasons explained in the footnote. When stars are plotted in the colour–magnitude diagram, they do not fall uniformly all over the diagram; most lie on 'the Main Sequence' – the name simply means that the majority of stars lie on this line.

The Main Sequence runs from hot, bright stars to faint, dim ones, and astronomers have discovered that the hot, bright ones are the more massive and the faint, dim ones are less massive. All stars on the Main Sequence, no matter how big or small, are called dwarves. The hottest ones (the most massive) are called blue dwarves, the coolest ones (the least massive) are called red dwarves; the Sun, which is of intermediate mass, and which has sat on the Main Sequence in equilibrium for 5000 million years, is a yellow dwarf.

Thus, I can describe the formation of stars in terms of the colour–magnitude diagram, rather than, as above, in terms of what is happening in physical terms.

Protostars form from the interstellar medium as cool, bright objects, moving quickly down the colour–magnitude diagram until they meet the Main Sequence. They stabilise at this point and sit on the Main Sequence for many millions or billions of years. If they are massive, they sit at the upper end of the Main Sequence. If they are not massive, they sit at the lower end. If they are too small, they never stabilise and fall out of the bottom of the colour–magnitude diagram. If they are too big they never stabilise and quickly move away from the top region. Since massive stars are less common than lighter ones, the upper reaches of the Main Sequence are less populated than the lower reaches.

When the hydrogen runs out

At first the stars are almost entirely made of hydrogen – this is the most common element in the Universe, having been made from protons and electrons created in the Big Bang. There is also a small percentage of helium, also created in the Big Bang. In addition the stars contain traces of heavier elements which have accumulated in the interstellar medium as a kind of pollution from generations of stars past.

But gradually this composition changes. Protons from the hydrogen of which the stars are principally made stick together to make helium in the nuclear reaction which is generating the energy of the star.

This increases the proportion of helium in the star material. It also, rather surprisingly, increases the proportion of nitrogen in the star material. This is because hydrogen burning in massive stars works by an efficient but complicated route called the CNO cycle. Essentially four hydrogen nuclei

Table 5. *The CNO cycle of hydrogen burning in massive stars*[a]

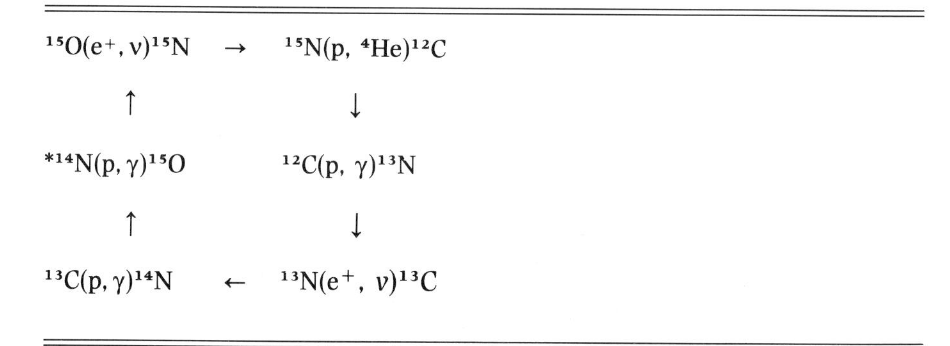

$^{15}O(e^+, \nu)^{15}N$	$\rightarrow$	$^{15}N(p, {}^4He)^{12}C$
$\uparrow$		$\downarrow$
$^{*14}N(p, \gamma)^{15}O$		$^{12}C(p, \gamma)^{13}N$
$\uparrow$		$\downarrow$
$^{13}C(p, \gamma)^{14}N$	$\leftarrow$	$^{13}N(e^+, \nu)^{13}C$

[a] The notation A(b, c)D means $A + b \rightarrow D + c$.
*This is the slowest part of the cycle

interact in succession with carbon, nitrogen and oxygen nuclei resulting in the creation of helium and the regeneration of the carbon, nitrogen and oxygen nuclei. The main cycle is given in Table 5.

The slowest part of the CNO cycle is the reaction marked by an asterisk in Table 5. It acts like a section of slow moving water in a series of rapid waterfalls. Material accumulates in a pool at the slowest moving point. Thus nearly all the carbon, nitrogen and oxygen, which are initially more or less equally abundant in the star material, convert to nitrogen as it builds up at that place in the cycle, and nitrogen becomes more abundant than the other two elements.

In a star like the Sun 500 million tonnes (5×10^8 tonnes) of hydrogen is changed to helium each second. The mass of the Sun is, however, 2000 million million million million (2×10^{27}) tonnes, and in the 5000 million years (1.6×10^{17} s) of its lifetime as a yellow dwarf, the Sun has converted only 4% ($5 \times 10^8 \times 1.6 \times 10^{17}/(2 \times 10^{27}) = 0.04$) of its mass to helium.

However, a more massive star (say 20 times more massive than the Sun) has a luminosity 60 000 times that of the Sun, and is consuming fuel proportionately more quickly. Such a star cannot sustain a loss like this for the same time as the Sun has existed without effect. For example, if it burns hydrogen at a constant rate, the star has been 30% converted to helium after only 12 million years (5000 million years $\times$ (30/4) $\times$ 20/60 000). This calculation takes no account of subtleties but serves to show that the hydrogen nuclear fuel in the centre of such a massive star runs out relatively quickly.

As a significant amount of the hydrogen runs out and is replaced by helium, the massive star re-adjusts its structure and its gravity distribution. Its inner

Fig. 16. In the Hertzsprung–Russell (H–R) diagram the luninosity of a star is plotted against its surface temperature. Here the luminosity is represented by the so-called absolute bolometric magnitude M_{bol}, and the temperature, T_{eff} is represented logarithmically. The evolutionary track of a star of 15 solar masses is drawn.

The star starts on the Main Sequence, its position in the lower left corner, where it has an absolute magnitude of −6 and a temperature of 30 000 K. It is a blue dwarf.

It slowly moves upwards by 0.8 mag (doubling in luminosity) at which point the hydrogen in the centre runs out. It is a blue giant at this stage.

The star moves rapidly at a magnitude of −6.8 across the diagram to cooler temperatures (to the right), and helium ignites in the centre of the star when its surface temperature is a cool 5000 K. It is then a red supergiant.

The star brightens by a further 0.6 mag and then moves rapidly to hot temperatures again. It begins carbon burning when its surface temperature is 20 000 K. It is at that time a blue supergiant.

The subsequent stages of nuclear burning progress very rapidly from this point, and the star does not have time to move much further in the H–R diagram before it explodes as a supernova. Sk−69 202 was near this stage just before it exploded (see next figure).

This diagram is simplified from calculations by Stan Woosley and his collaborators for a 15 solar mass star with restricted convection and a composition with heavy metal abundances $\frac{1}{4}$ of those of our Sun, intended to represent the average of the LMC.

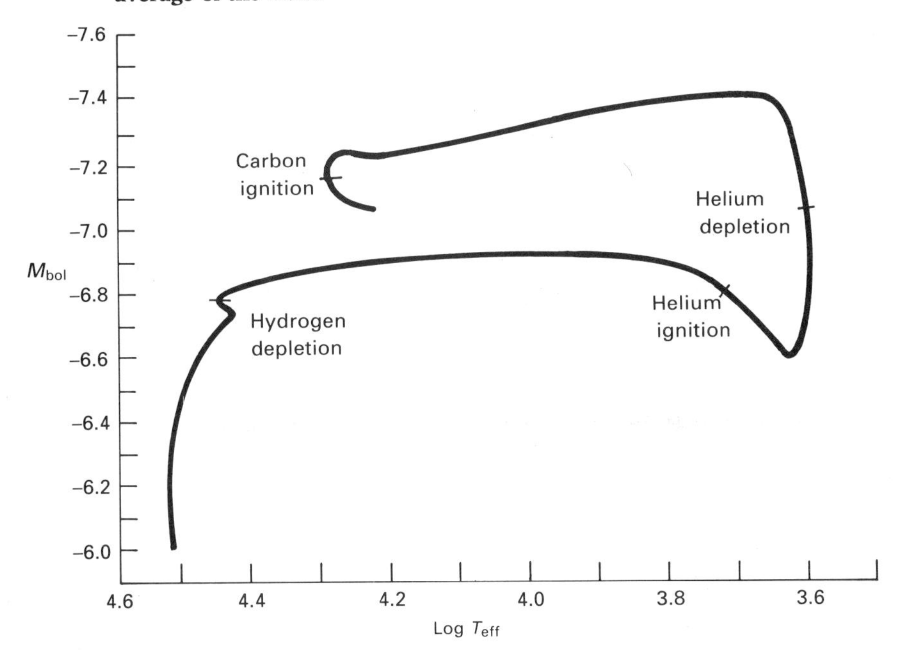

regions contract, and the energy released as they do, raises the central temperature enough to ignite a new nuclear fuel – the helium which has been made by hydrogen burning. Three helium nuclei combine to make a carbon nucleus in this nuclear reaction.

The star changes its luminosity and surface temperature in response to the changes of structure. It moves off the Main Sequence, lifting up and increasing in luminosity as the hydrogen is used up. When a pure helium core of about 6 solar masses has formed, the star moves to the right of the Main Sequence, and (although its central regions are hotter) its surface becomes cooler. As it lifts off the Main Sequence it becomes first a blue giant, then a blue supergiant, and then a red supergiant.

Because of this behaviour, a group of stars which are all the same age (as, for example, a group of stars in a cluster which all formed from the collapse and fragmentation of the same gas cloud) form a very distinctive pattern in the colour–magnitude diagram. The more massive ones have exhausted their hydrogen fuel and have lifted off the Main Sequence, but the less massive ones have not and still remain in equilibrium on it.

The mass of Sk − 69 202

It is possible to plot the colour–magnitude diagram of the three stars (Stars 1, 2 and 3) in the region of Sk − 69 202. The assumption is that these three stars are related one to another (van den Bergh 1987). It is not a very well-defined colour–magnitude diagram, since there are only three points in it, but we know overall what the diagram looks like anyway.

Stars 2 and 3 lie on the Main Sequence – they are blue dwarves. Star 1, Sk − 69 202, did not – it was a blue supergiant. The three stars were presumably formed at the same time as part of a triple system. Star 1 must have been the most massive – it had already exhausted its hydrogen fuel. Stars 2 and 3 are less massive and still are burning hydrogen. Star 2 is 10 times the mass of the Sun, and so Star 1, Sk − 69 202, must have been more massive than this, say 15–20 solar masses. The consensus is that Sk − 69 202 was about 17 solar masses.

Type II supernovae as the explosions of massive stars

It is consistent with what astronomers believe for SN 1987A to have been a Type II supernova caused by the explosion of a massive star like Sk − 69 202. Astronomers believe that, in general, Type II supernovae represent the explosions of stars whose masses are in excess of about 5–7 times the mass of the Sun. The way that they come to this conclusion is in a chain of deduction as follows.

First they connect the occurrence of Type IIs in the spiral arms of galaxies with the hot, bright stars which inhabit these regions. Then they use the colour–magnitude diagram to identify hot, bright stars with the massive stars at the upper end of the Main Sequence. They conjecture that all stars whose mass is over a certain minimum mass will explode as supernovae. They then match the rate at which supernovae occur with the rate at which stars leave the Main Sequence. They work down from the top of the Main Sequence, totalling the rate at which stars leave it, until they have reached a sufficiently large number. If all stars above 5–7 solar masses explode as supernovae, then this generates the right number of supernovae. And it fits with the discovery that Sk – 69 202 was about 17 solar masses.

Fig. 17. Stars of all kinds and all ages populate the H–R diagram, stopped in their long lived tracks by the snapshot which constitutes the brief history of astrophysics. The diagram for the brighter stars of the LMC shows blue supergiants (including, arrowed, Sk–69 202) and red supergiants in a group at the right. The gap between blue and red supergiants is real, because stars sprint across this region and are seldom caught in the act. The dearth of fainter stars in the lower part of the diagram, however, just represents the difficulty of measuring them in another galaxy, even if it is the nearest one. The slanting line to the left is the Main Sequence and tracks from it represent calculations of the path of stars which have the masses labelled.

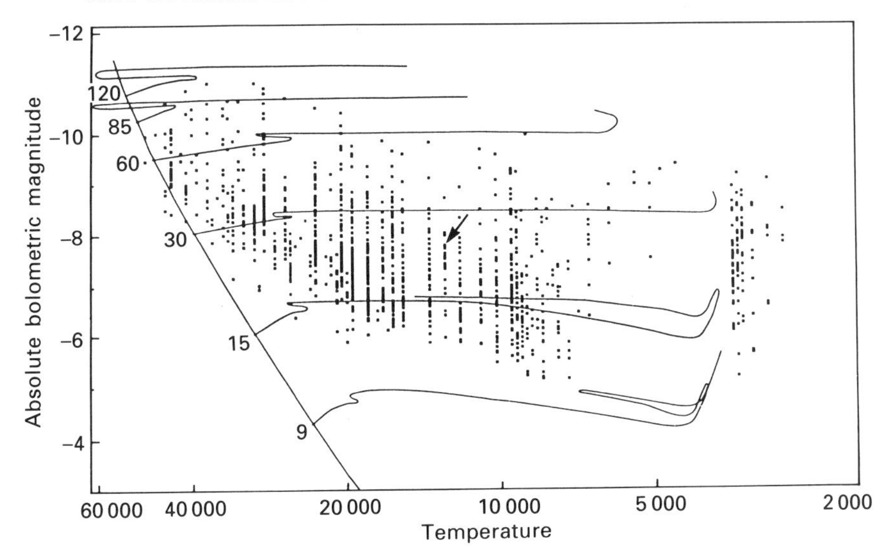

Massive stars in later life

I left the description of the evolution of massive stars like Sk – 69 202 at the stage where it had built itself a helium core of 6 solar masses, which was cocooned in an envelope of 11 solar masses of hydrogen. The helium was burning at the centre to make carbon and oxygen. In later life Sk – 69 202 even ran out of helium as a nuclear fuel at its centre. It began to burn the carbon to make neon, sodium and magnesium. The helium burning zone moved outwards. The star gradually built up an onion-like structure with layers of different elements.

Its inner regions were made of concentric zones of alpha-nuclides (nuclides made of assemblages of four nucleons) (Hillebrandt, Hoflich, Truran and Weiss 1987; Woosley *et al.* 1987; Arnett 1988). The outer layer remained principally hydrogen; then came a layer principally of helium; the next region was principally carbon; and the innermost core was made of the heavier alpha-nuclides like oxygen, neon, magnesium etc.

At the boundaries between the layers of the onion, the lighter element was being transmuted into the heavier by thermonuclear fusion. Thus there was a shell of energy generation where four protons were being converted to alpha particles (hydrogen burning), another where alphas were being fused to make ^{12}C and ^{16}O nuclei (helium burning), and another where alphas were being added to ^{12}C to make neon, sodium and magnesium (carbon burning), etc.

When each nuclear fuel in the centre of the star ran out and the pressure inside the star altered, the star re-adjusted its structure and its gravity distribution in order to maintain a balance between pressure and gravity.

In making its adjustment, the star's inner regions contracted, heated up and raised the central temperature enough to ignite the next stage in the nuclear chain. Because the conditions inside the star had altered, with the core hotter, and the composition of the layers of the star changed by the nuclear fusion, the outer parts of the star also altered – and these are the parts which show to the rest of the Universe. Thus, the star changed yet further its luminosity and surface temperature. It did this in a quite complicated way in response to the internal changes of structure, and it moved in a complex track across the colour–magnitude diagram, from blue to red, and red to blue, back and forth.

The situation was even more complicated by the likelihood that a hot, bright star like Sk – 69 202 would lose considerable amounts of its mass in moving about the colour–magnitude diagram. It was a large star when it was a blue giant and a red supergiant, and its surface gravity was relatively small. At the same time its luminosity was high, so it gave lots of energy to its atmospheric material. The net result was that considerable amounts of mass were pushed off Sk – 69 202 in the course of its life (Chapter 11).

Mass loss seems to be quite a general phenomenon in stars and the lost mass often manifests itself as emission lines. When they are noticed in the spectrum of hot stars, the stars are known as 'Of' type. Red supergiants often have emission lines and other indications that they are losing mass. Perhaps half or more of the star is lost in such a 'stellar wind', and surrounds the star in a circumstellar cocoon. So much of the outer parts of the star may be lost that it is possible to see the insides of the star and view the nuclear processed material as it is blown off in the stellar wind. Such stars have spectra with strong nitrogen or carbon emission lines and are called Wolf–Rayet (WR) stars, after the two astronomers who identified them as a class.

When a massive star grows large in the course of its evolution, its wind certainly and perhaps also the star's outer layers may reach any other nearby

Fig. 18. In the evolution of a 20 solar mass star, about half the mass is lost in a stellar wind, while an onion-like structure of concentric shells of different elements builds up within the star. Three sectors represent portions of the star at three different stages in a calculation of the changes in a star similar to Sk – 69 202. The first sector is drawn at age about 5 million years, the second at 10 million years and the third at 11 million years. The radial scale is distorted in these figures, with the inner dense regions enlarged relative to the more rarefied outer parts, so that a linear distance represents the mass enclosed within that distance – in reality the red supergiant is enormously larger than the blue dwarf or even the blue supergiant. By the final stage, there is at the surface of the star a thin hydrogen layer (contaminated by helium from previous stages). There is a shell of hydrogen burning inside the surface layer, and inside that about 3 solar masses of helium, on the inner edge of which helium burning is occurring. Seven solar masses of the star are a carbon–oxygen–neon mixture with a central carbon–burning core. Based on calculations by A. Maeder.

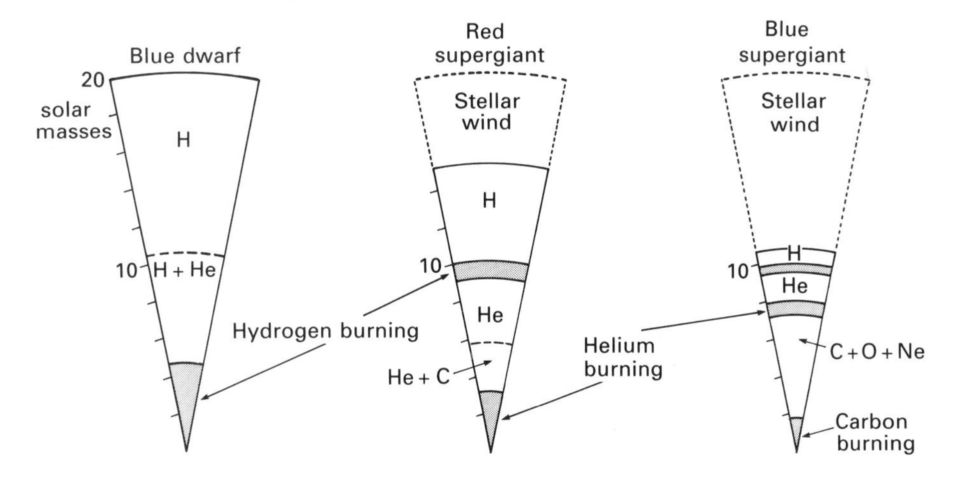

star, any companion. The gravitational pull of such a companion will also affect how much material is lost. It is very common for stars to have companions, and especially common for massive stars. The effect of the companion will depend on the strength of its pull – and this depends on how massive the companion is and how near.

In summary, the outer parts of a star are not well coupled to what is happening in its deep interior – there is no quick one-to-one connection between the core and the outer layers. If the core of a star changes, it takes time for this change to affect the outer parts. This means that when the core of a star has reached a particular stage, the outer parts (the parts which we see and by which we classify the star) can be quite different from each other. The supernova explosion is triggered by something happening in the core – and the outer parts only loosely indicate that the explosion is imminent.

This range of possibilities for the evolution of massive stars has quite an effect in varying the circumstances of a Type II supernova explosion. In the same way, a bomb explosion may look quite different if the bomb is underground or in air. Even taking into account that the bomb might be manufactured in a factory the factory is being altered while the bomb is being readied. At the time when the bomb goes off, the factory might still be being enlarged, or dismantled; it might have been constructed of reinforced concrete or flimsy plywood. The explosion of the bomb will have different consequences in all these cases.

The upshot is that we can expect quite a range of properties of Type II supernovae. Indeed, to take one such property – the brightness – Type IIs show quite a dispersion of luminosity compared to Type Is. The brightness of Type Is is so well defined that they are thought to be standard candles – so much the same brightness that they can be used in cosmology to determine the distances of their host galaxies. But the Type IIs cannot be used this way – they are too heterogeneous.

The spectra of Type IIs may also be very different: in particular if a supernova explodes after the hydrogen on the outer parts has been lost, then there may be no hydrogen lines in the optical spectrum and the supernova will not look like a Type II supernova at all – it will masquerade as a Type I. Some astronomers conjecture that this is the explanation for the Type Ib supernovae: they are perhaps the explosions of the stripped cores of massive stars.

A classification (Table 6) of the various kinds of progenitors of massive star supernovae has been made by Maeder (1987). He emphasises that it is tentative and could be different in different galaxies or different places within the same galaxy. It takes no account of whether the stars are members of binary systems.

Table 6. *The different kinds of progenitors of massive star supernovae*

Progenitors	Mass range (solar masses)	Evolutionary sequence of the progenitors[a]
Wolf–Rayet stars	>40	*Always blue stars* O-stars→Of-stars→blue supergiants→ Wolf–Rayet stars→supernovae!
	20–40	*Blue → red → blue* O-stars → blue supergiants → red supergiants → blue supergiants → Wolf–Rayet stars → supernovae!
blue supergiants	near 20	*Blue → red → blue* O-stars → blue supergiants → red supergiants → blue supergiants → supernovae!
yellow or red supergiants	<20	*Blue→red→blue→red or yellow* O-stars→blue supergiants→red supergiants→ blue supergiants→red or yellow supergiants→ supernovae!
red supergiants	7–18	*Blue → red* O-stars → blue supergiants → red supergiants → supernovae!

[a] O-stars are bright, hot stars. The letter O represents the appearance of the spectrum of such stars, and is a linguistic fossil from the period of astronomy when the study of stellar spectra was based on empirical classification schemes, with only a broad idea of what the spectra might mean. Later astrophysical investigations showed that O-stars are the massive stars, which populate the spiral arms of galaxies, and which give rise to Type II supernovae. In the same days of past history, some of the fine details in O-star spectra were represented by additional letters, and an Of-star is an O-star with emission lines in its spectrum; the emission lines represent an atmosphere surrounding the star.

Sk − 69 202 — 'blue when it blew'

To what state exactly had Sk − 69 202 evolved before its explosion? In Figure 17 (p. 65) the position of Sk − 69 202 is plotted on a colour-magnitude diagram of the stars in the LMC. To the right is a bunch of stars which are red supergiants. To the left are the blue supergiants. Sk − 69 202 lies midway between the groups, on the red side of the blue supergiants. Superimposed on the diagram are lines representing attempts to calculate the evolutionary paths of LMC stars. Sk − 69 202 is about 17 solar masses, but it is not clear from the position of the star whether it was moving from blue to red, or red to blue.

Fig. 19. A detailed diagram, with three successive enlargements of the middle regions, shows another calculation of the structure of a blue supergiant star similar to Sk – 69 202 at the moment of core collapse. Here the calculation, by Stan Woosley and collaborators, assumes that 4 solar masses of the 20 solar mass star has been lost by the moment of core collapse. The radial dimension, given at the top of each diagram, is to scale and the central region of the star is enlarged from drawing to drawing by factors of × 600, × 10 and × 10 respectively. The density (g/cm³) in the various zones of the star is labelled along the radius to the left of each diagram, the temperature (millions of degrees Kelvin) is labelled to the right, and the mass of each zone (solar masses) is labelled to the bottom. The iron core in the centre of the last drawing is less than the diameter of a human hair when viewed on the scale of the first drawing, and is 1/8000 million million of the volume of the blue supergiant star, although it contains about 1/10 of its mass.

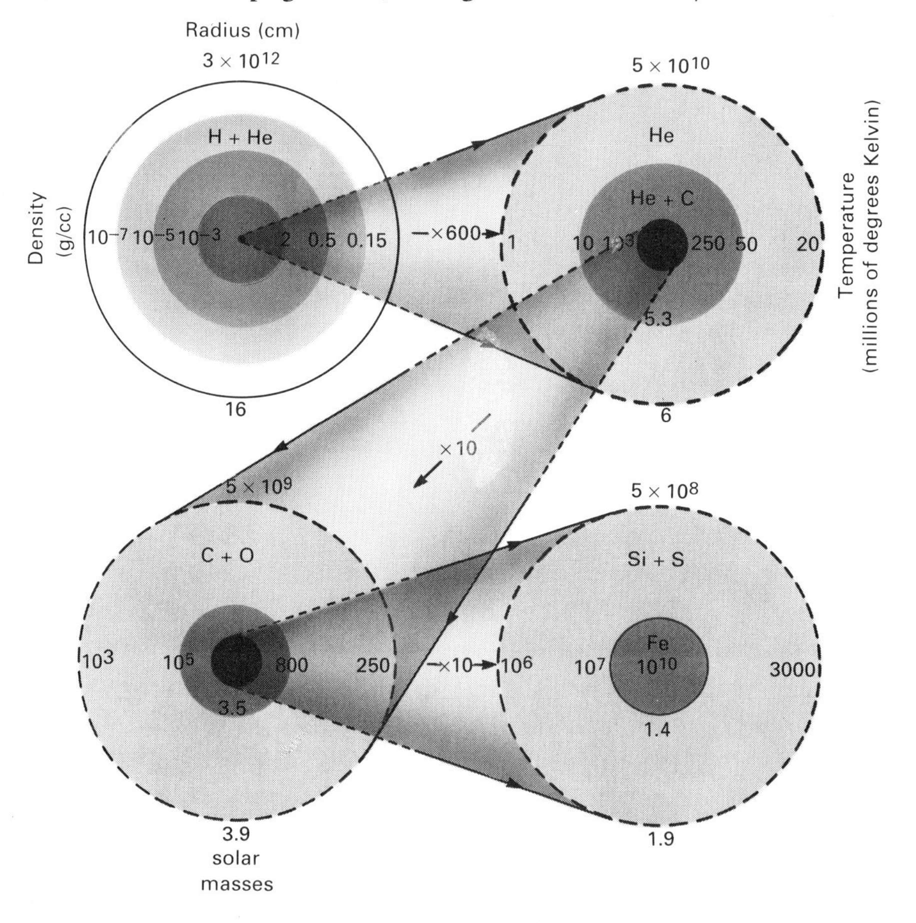

'It was blue when it blew, but was it always blue?' is the question as jauntily framed by Kirshner.

It seems likely (Maeder 1987) that after moving off the Main Sequence Sk − 69 202 became a red supergiant, where it remained for about 8% of the time that it had spent on the Main Sequence, and then it returned to become a blue supergiant just before it exploded.

The latter stages of the chain of the thermonuclear fusion processes occur very quickly, after a slow start at the beginning of the chain. The tempo of the star's life accelerates dramatically.

We can express what happened to Sk − 69 202 in the LMC in terms of what was happening on Earth when its radiation arrived.

The star was forming nearly 20 million years ago during the Miocene epoch of geological history. It burnt hydrogen for some 15 million years, throughout the Miocene and Pliocene epochs, while the higher mammals were evolving.

Helium burning started at the end of the Pliocene, at the time when the first ape-men were appearing on the Earth, and lasted the million years it took to evolve the human race.

Sk − 69 202 began carbon burning some time in the slow dawn of human history, somewhere near the time of the major land-migrations into Australia and the Americas, the creation of the cave-paintings of Lascaux or the establishment of great archaeological monuments such as Stonehenge and the Pyramids; Sk − 69 202 was at this stage for the several thousand years of the whole of history.

By 1980, Sk − 69 202 was neon and oxygen burning.

On about 1987 February 20, Sk − 69 202 began silicon burning. This stage lasted a few days and produced the iron core.

On 1987 February 23 at 07 : 35 UT the supernova explosion ended the star's life.

6
The explosion of a supernova

End of the fusion chain

There comes a point in heavy enough stars when the chain of fusion reactions and the development of successive stages in the shell structure cease. The successive fusion reactions produce a chain of nuclides with the release of energy:

$$H\ {}^{4}He\ {}^{12}C\ {}^{16}O\ {}^{20}Ne\ {}^{24}Mg\ {}^{28}Si\ {}^{30}Si\ {}^{32}S\ {}^{34}S\ {}^{40}A\ {}^{49}Ca\ {}^{48}Ti\ {}^{52}Cr\ {}^{54}Fe\ {}^{56}Fe.$$

But around iron the chain breaks.

Beyond the ${}^{56}Fe$ nucleus, in the stages to ${}^{60}Ni$, ${}^{64}Zn$ . . . the successive links in the fusion chain require the *input* of energy; they do not release it. The technical term for this is that the reactions are endothermic. If the star goes on any further in the nuclear fusion chain it gets colder rather than hotter.

The pressure in the star begins to die away. But the relentless force of gravity still remains. The star contracts. The force of gravity pushes the nuclei and the electrons together and they merge by the process of electron capture, each nucleus resulting in several neutrons. Because the electrons effectively disappear from the core, the pressure dies away even further.

Moreover, the nuclei are torn apart by the high energy radiation in the hot core (photodisintegration). This undoes the work which went into building them during the fusion chain in the previous life of the star. In making the nuclei, energy was released into starlight – in tearing them apart, energy is absorbed from the core. This too progressively decreases the pressure inside the star.

So no processes occur which turn up the pressure in response to the contraction of the core – on the contrary, the outward pressure is diminished by the core's contraction, and the weight of the core becomes ever more dominant. The core collapses further, increasing the speed of its fall. It becomes a neutron star, made of neutron-material.

Core collapse

The core of Sk – 69 202 just before collapse was of mass $M = 1.4$ solar masses and radius about $R = 6000$ km. It was made of nuclei and free electrons in a hot plasma. It density was

$$D = M/(4/3)\pi R^3 = 3 \times 10^{33}/(4/3)\pi(6 \times 10^8)^3 = 3 \times 10^6 \text{ g/cm}^3.$$

The core forms a neutron star. Neutron-material is of density $d = 1 \times 10^{15}$ g/cm³. The density of the star has thus become 100 million tonnes per cubic centimetre. The radius of the neutron star has shrunk to about $r = 10$ km.

The timescale for the collapse is amazingly short. The actual behaviour as the core collapses is complicated by the pressure forces which resist the free fall, but we can neglect these forces and make an estimate of the timescale by thinking about the gravitational potential energy which is released as the star shrinks from radius R to radius r while it increases its speed to v in time t.

Just like any object in a gravitational field, say a pen held above a desk, the material of the core in its own gravitational attraction has energy which can potentially be turned into some other form – energy of motion if the pen is dropped or if the core collapses.

The gravitational potential energy of the material of the core depends on the mass M of the core, its radius (the smaller the core gets the more energy is available to accelerate the core) and a physical constant of gravitation called G. The gravitational energy of the core is of order GM^2/R, and the gravitational energy of the neutron star is of order GM^2/r. Since $R \gg r$, the energy released initially into kinetic energy is $GM^2/r = \frac{1}{2}Mv^2$. The time to drop a distance R is

$$t = R/v = R/(GM/r)^{\frac{1}{2}}$$
$$= 1/(GM/R^2r)^{\frac{1}{2}} = 1/(G(R/r)M/R^3)^{\frac{1}{2}}.$$

Since $D = M/(4\pi/3)R^3$

$$t = (r/RGD)^{\frac{1}{2}}.$$

(This is a very rough and ready calculation which ignores the acceleration of the core from a standstill, the fact that it is not uniformly filled, and neglects the pressure forces which do resist the free fall, so it is not worth being accurate about factors like 0.5 or $4\pi/3$!).

For $r = 10\,000$ m. $R = 6$ million m, $D = 3 \times 10^9$ kg/m^3, and $G = 6.7 \times 10^{-11}$ N m^2/kg^2, we find $t = 0.1$ s.

The time, t, for the collapse of the core is less than a second!

The quantity $(GM^2/r - GM^2/R)$ is the amount of gravitational potential energy which is released. Since r is much less than R the second term is negligible. GM^2/r is called the binding energy of the neutron star and it is about 6×10^{46} J.

Energy release

The binding energy of a neutron star is a lot of energy. This is not merely equivalent to the statement that number like 6×10^{46} is very large number, because if I express the binding energy in ergs (10^7 erg $= 1$ J) the energy might look even larger – it is 6×10^{53} erg, a bigger number which expresses the same amount of energy. So what does 'large' mean in this context? To put the energy released in the core collapse into its true perspective, compare it with the energy which the Sun radiates. The Sun's luminosity at the moment is $L_\odot = 4 \times 10^{26}$ J/s. The Sun has lasted for 5000 million years and will last as long again, and so its lifetime is 3×10^{17} s. Over its lifetime it will radiate 10^{44} J. The core collapse releases more than 100 times more than this.

Put it another way. The gravitational potential energy released by the core collapse can be converted to the mass-equivalent by Einstein's formula $E = mc^2$, where c is the velocity of light. If the mass m is totally annihilated then an energy E is released; conversely, the energy E is equivalent to the annihilation of the mass m. At the present time 4 million tonnes of the Sun's mass are annihilated every second. The Sun will shine for 10 billion years in total, continuing to radiate away this amount of its mass every second. And the total mass which it will lose in this time is still 100 times smaller than the mass annihilated in the supernova's core collapse, which sums in total to about 0.1 of the mass of the Sun.

Bear in mind that this energy is released in less than one second; most of it is radiated away, as we shall see in the next chapter, in the form of neutrinos. The luminosity of the supernova is thus something like 10^{47} J/s. This is 10^{21} times the luminosity of the Sun, $L_\odot$. Compare this with the fact that there are 10^{10} galaxies in the Universe each containing 10^{11} stars, and that, since the Sun is an average star, the luminosity of the whole Universe is $10^{21} L_\odot$. The luminosity of the supernova is thus the same as that of the whole Universe.

The binding energy is converted into kinetic energy as the core collapses, just as the potential energy of the pen held above the desk top is converted into kinetic energy when it falls. And just as the kinetic energy of the pen is converted into heat when it hits the desk top and comes to a halt, so the kinetic energy of the collapsing core is converted into heat when it stops its inward fall and becomes a neutron star. The heat warms the collapsing core material to temperatures of order $T = 10^{10}$ K.

The collapse continues

When the central core has collapsed it removes the support from the layers surrounding it; these also fall inwards. Because they are further removed from the dense centre they experience a weaker gravitational pull and they

Fig. 20. Four glimpses of the core, mantle and outer envelope of the star (not to scale), as the core collapses. At first the inner core collapses within the mantle, forming a neutron star, and generating neutrinos. The mantle follows the collapse of the core and bounces on the hard surface of the neutron star. A shock wave propagates after the neutrinos, through the outer part of the star, causing the envelope to explode.

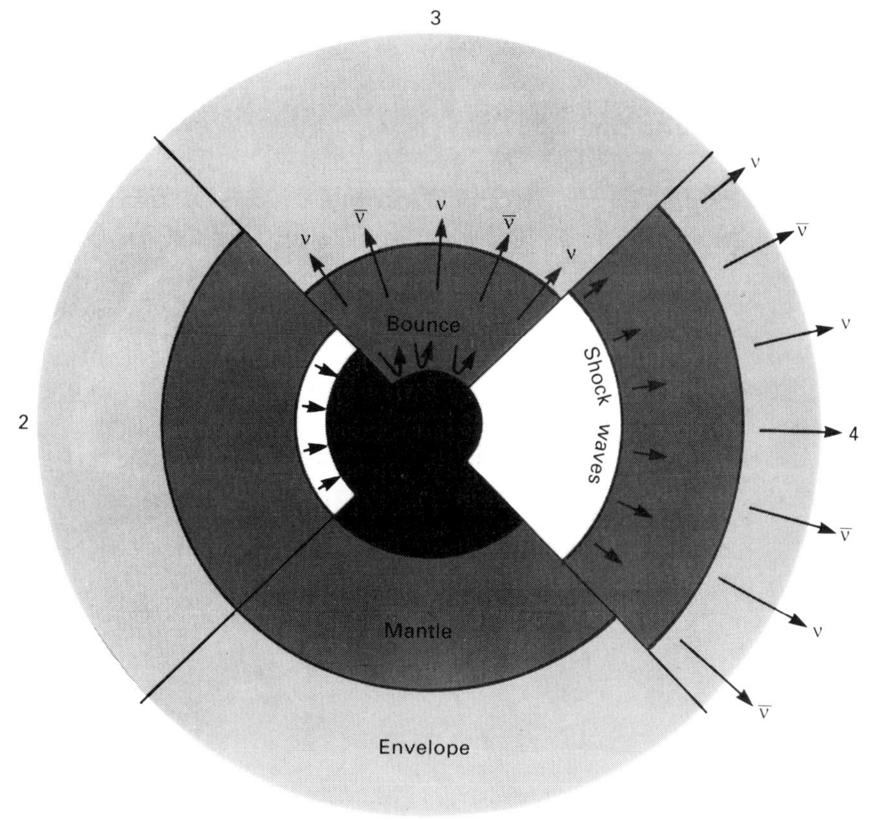

have further to travel to the centre than the core. They move slowly at first. When they get to the centre of the star they find that the neutron star has already formed. They suddenly encounter its hard, rigid surface and they collide with it abruptly.

The layers of material which surrounded the iron core 'bounce'. In addition, heat energy, which the formation of the neutron star released, has been travelling upwards against the inward fall of the outer layers, pushing against them. Their inward collapse is reversed. The outer layers of the star explode outwards, pushing strongly on the layers above, aiding the push of the flow of energy.

The ultraviolet pulse

The heat and the abrupt collision of the star material on the hard surface of the neutron star create a shock wave which travels outwards through the star. The shock travels quickly through the star: it is a supersonic† wave. In fact, the word 'shock' means that the wave is supersonic. Any pressure disturbance (sound) created by the shock wave travels slower than the shock itself, so the layers of the star above the shock are unwarned about the shock wave coming and make no attempt to get out of the way of the shock. They suffer the full impact of the shock wave collision.

The surface of the star remains unaware of the shock travelling upwards from below until the moment that the shock arrives – until this moment, an observer looking at the surface would not have been aware that the core had collapsed. The shock wave breaks out from the surface of the star about one hour after the core collapses. So, in the morning of 1987 February 23, the surface of Sk – 69 202 was suddenly heated by the shock to temperatures of $(0.5 \text{ to } 1) \times 10^6$ K (Ogelman *et al.* 1987).

The energy which is radiated by the flash, mostly in the form of ultraviolet photons, can be found by scaling the luminosity of Sk – 69 202. Its luminosity just before the shock reached the surface was $L_* = 40\,000$ times the luminosity of the Sun and its surface temperature was 28 000 K. Moments afterwards, it was the same size as it was before, but the temperature was 1×10^6 K. Thus its luminosity was

$$L_{\text{shock}} = \frac{(1 \times 10^6)^4}{(2.8 \times 10^4)^4} L_*$$
$$= 1 \times 10^6 L_*$$
$$= 40 \text{ thousand million solar luminosities.}$$

† Supersonic means, of course, travelling faster than the speed of sound. It is implied that we are referring to sound travelling in the star material.

Although most of the radiation was in the form of ultraviolet photons, the light from Sk – 69 202 also suddenly increased, perhaps 30-fold. Sk – 69 202 thus virtually instantaneously brightened from magnitude 12 to 9.

Radiating at such an immense luminosity, the surface rapidly cools. The high temperatures fade away quickly, so that there is an ultraviolet flash of duration several minutes.

If there was no interstellar material, nor atmosphere surrounding the Earth, then Sk – 69 202 would have brightened in the ultraviolet to be the brightest star in the sky. But wavelengths longer than 40 Å are cut off by the voyage of the ultraviolet radiation through the gas of interstellar space. The total amount of ultraviolet energy which the supernova would have delivered to the top of the atmosphere of our Earth was between 0.01 and 1 erg/cm^2.

The radiation would have covered the wavelength region from about 12 Å to about 40 Å. Soft X-rays like this penetrate into the Earth's atmosphere down to altitudes of 110–125 km (the E-layer). They ionise the air molecules, leading to an enhancement of the electron concentration in the atmosphere. The

Fig. 21. Map of the world showing the terminator, the division between day and night at the moment of the break-out of the shock wave from the surface of Sk – 69 202. The LMC was overhead in Antarctica at longitude 150°, latitude – 80° at that moment. The dotted lines are contours of the peak enhancement of the density of electrons in the Earth's atmosphere caused by the ultraviolet flash from the shock break-out, in units of electrons per cubic centimetre. Calculations and figure by H. Ogelman and collaborators.

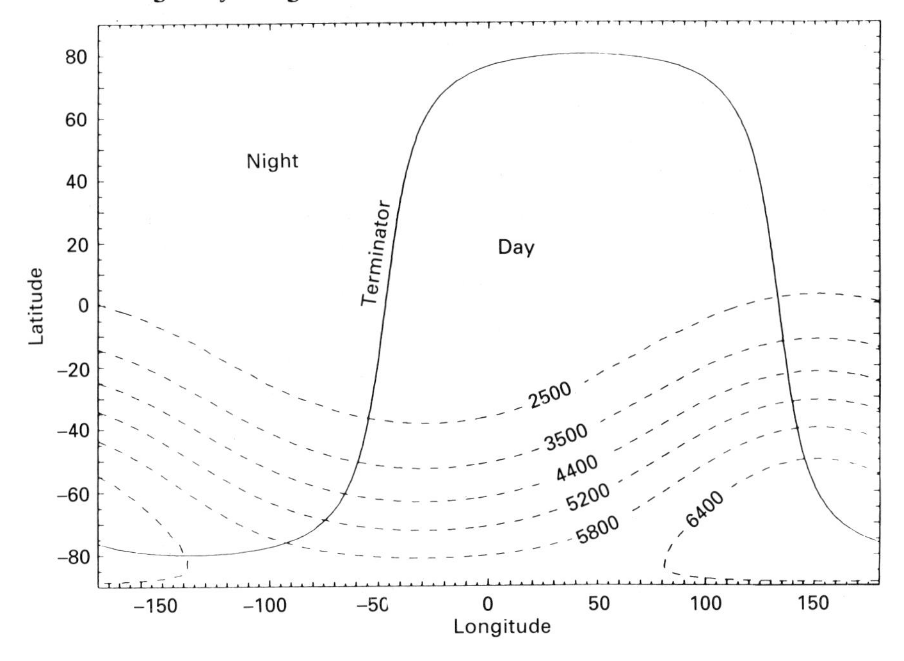

enhancement would have lasted for the several minute duration of the ultraviolet flash and decayed thereafter in a few minutes as the molecules recombined – up to a quarter of an hour in total.

For places on Earth where the supernova, at the time of the shock break-out, was directly overhead at night, the electron density of the E-layer would have increased by several times its natural value, leading to measurable radiopropagation anomalies, and fluctuations of the magnetic field of the Earth. Unfortunately, at the time that the shock broke out of the surface of Sk – 69 202, the supernova was in fact overhead in Antarctica, south of New Zealand, before nightfall. There are few geophysical measuring stations in this region and the effects would not have been as pronounced in the late afternoon as during the night-time. These effects therefore remain as predictions (Ogelman *et al.* 1987) and not as observations.

Fig. 22. Height profile of the electron density in the Earth's atmosphere at night-time (crosses) shows a peak of 4000 electrons per cubic centimetre at an altitude of 120 km (the E-region). The solid lines are calculations of the profiles of electron density produced by a supernova in the LMC with the temperature of its shock break-out 1×10^6 K degrees and total energy in the burst of 10^{48} erg. Curve (*a*) represents a supernova seen overhead and (*b*) a supernova seen at 45° altitude. Such a supernova seen overhead doubles the night-time peak of electron density in the E-region.

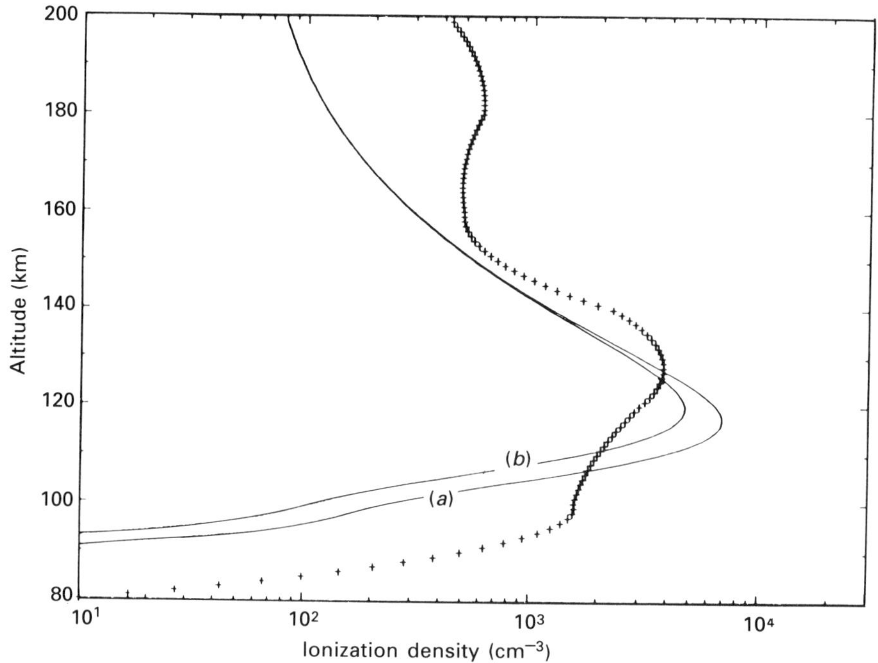

Expansion of the supernova

After the break-out of the shock from its surface, the various layers of Sk – 69 202 then began to expand significantly. The outer surface expanded with speeds of up to 0.1 the speed of light, or more. To first order, the star expanded homologously, growing in size uniformly but retaining its shape – it is as if a zoom lens was focused on it and its image was enlarged without distortion. The radius of the star increased rapidly – by a factor of 5 in the first hour of the explosion. In ten hours the star enlarged to the size of the Earth's orbit around the Sun. The star's surface area increased by 25-fold in the first hour and 2500-fold in the first ten hours. The increase in surface area more than compensated for the cooling of the star after the flash of ultraviolet heating at shock break-out. Thus the star continued to brighten.

In this way, in a matter of hours, Sk – 69 202 brightened to a considerable magnitude where no bright star was particularly noticeable before.

Sk – 69 202 became a supernova.

7

Neutrinos from the inferno: the core collapse of SN 1987A

New particles for radioactive decay

Neutrinos are remarkable particles, with a remarkable history. They were 'invented' before they were discovered – a clever scientist realised that they must exist, although they had never been seen. The story goes like this.

In the 1920s, physicists who were studying the decay of the radioactive nuclei of some elements were faced with the problem of some missing energy. There were different kinds of radioactivity one of which was labelled beta. In beta decay, a neutron (embedded in a nucleus) produces a proton and an electron:

$$n \rightarrow p + e^-.$$

The neutron has zero electric charge, while the proton's electric charge is $+1$ (in certain units of electric charge) and the electron's is -1, and so this particular quantity in the transformation balances, with zero electric charge before and afterwards.

Electric charge conservation was true:
charge of source = charge of products
$0 = +1 + -1.$

But, it was found, energy was not equal on both sides of the equation.

Energy conservation was not true:
energy of source ≠ energy of products
energy of neutron ≠ energy of proton + energy of electron.

The energy, on the right hand side of the equation, of the proton and the electron was always less than that of the neutron, on the left, and energy was not balanced on both sides of the equation. This was true even when you took account of the difference in masses of the particles and recognised that mass is equivalent to energy ($E = mc^2$) and has to be taken into account.

Energy conservation, including mass–energy equivalence, was not true:

rest mass energy + kinetic energy of neutron
≠ rest mass energy + kinetic energy of proton
+ rest mass energy + kinetic energy of electron.

In fact, the measurements of the energy were not the crucial discovery: there was a general statement which could be made about the energy which made it plain that something was wrong. It was observed that the electron which had been created could have any energy in a broad spectrum between zero and a certain maximum.

The deduction was quite simple. Protons, neutrons and electrons all have a given rest mass and therefore a given rest energy. But the neutrons and protons have no kinetic energy because they remain bound up in the stationary parent nucleus. The nucleus is stationary because it is fixed into the crystal lattice of the material which is undergoing radioactive decay. The electrons on the other hand were ejected at high speeds – they were, in fact, the beta-rays which had drawn attention to the radioactivity in the first place. If energy conservation including mass–energy equivalence was correct, the neutron's energy must be shared out between the fixed rest-mass energies of the proton and the electron, and the electron's kinetic energy. The kinetic energy of electron would be equal to the difference between the rest-mass energies of the neutron with which beta decay started and the sum of the rest-mass energies of the proton and the electron with which it ended. According to this argument, the electron's kinetic energy must be a unique value, since the masses of the three particles in beta decay were all fixed. In effect, this logic described the shape of the distribution of the kinetic energy of the electron – its energy spectrum would be a spectral line.

In fact the energy of the electrons formed a continuum of values over a range between zero and a certain maximum. This could only happen if each electron's kinetic energy was being shared between it and another particle[†].

† The argument given ignores the possibility that the neutron and/or proton can be in excited levels of the nucleus. This means that the electron can have a number of selected energies, and its kinetic energy spectrum can be several spectral lines, not just one. There is still a great difference between a line spectrum and a continuum.

There was another problem: the electron, proton and neutron all had angular momentum values (spin). Spin ought to be conserved, just like the energy or charge, but was not.

Spin conservation was not true:
spin of source $\neq$ spin of products
spin of neutron $\neq$ spin of proton + spin of electron.

All three particles have spin $\frac{1}{2}$ (in certain units of angular momentum). In the logic of quantum mechanics, spin $\frac{1}{2}$ particles cannot spin in any orientation they choose; they can only spin parallel or anti-parallel to the measurement direction (i.e. 'up' or 'down'). Supposing we measure the spin of the neutron and it lies in the direction which we call 'up', with value $\frac{1}{2}$. The proton and the electron can have spins each of which lies 'up' or 'down'. If both are 'up', the total spin of the products is $\frac{1}{2}+\frac{1}{2}=1$ ('up'). This is not equal to the neutron's spin of $\frac{1}{2}$ 'up'):

$\frac{1}{2}=\frac{1}{2}+\frac{1}{2}$ is not true!

If both are 'down', the total spin is $-\frac{1}{2}-\frac{1}{2}=-1$ ('down'). This is not $\frac{1}{2}$ 'up' either.

$\frac{1}{2}=-\frac{1}{2}-\frac{1}{2}$ is not true either!

If one is 'up' and one is 'down', the total spin is $\frac{1}{2}-\frac{1}{2}=0$. So, finally, nor is this $\frac{1}{2}$ 'up' like the original neutron.

$\frac{1}{2}=\frac{1}{2}-\frac{1}{2}$ is the third combination and is not true.

There are no other combinations and, if the neutron's spin is reversed to its other possible value of $-\frac{1}{2}$, the argument simply repeats. Since none of the possible combinations adds up to spin $\frac{1}{2}$ 'up', spin cannot be conserved in this combination of products, just as energy is not conserved.

This was a second argument that a third particle had been created in the products of neutron decay.

These considerations led the Austrian physicist Wolfgang Pauli in 1930 to 'invent' an unseen, new particle to make the conservation laws work. The reaction for beta decay now reads:

$$n \rightarrow p + e^- + X.$$

X was the new particle. It had to be neutral (so as not to disturb one of the conservation laws which *did* work in the old understanding of beta decay – the conservation of charge). Of course it carried energy, and Pauli postulated that its spin was $\frac{1}{2}$, like the neutron, proton and electron. Then if the total spin of the proton and electron was $+1$ 'up', its spin could be $+\frac{1}{2}$ 'down', and the sum of all three would be $\frac{1}{2}$ 'up', just like the neutron's spin:

$\frac{1}{2}=\frac{1}{2}+\frac{1}{2}-\frac{1}{2}$ is true.

If the total spin of proton and electron summed to zero, then the new particle's spin could be $+\frac{1}{2}$ 'up', and again the sum of all three spins would balance the neutron's spin.

Fig. 23. Neutrons, protons and electrons have spin $\frac{1}{2}$ which can point up or down. If a neutron decays only to a proton and electron, the spins of the proton and electron never properly add to the same spin as the neutron had in the first place. But if a neutron decays to a proton, electron and a third particle with spin $\frac{1}{2}$, then, while there are some decay modes which do not add up properly, there are three decay modes in which the spins do add up

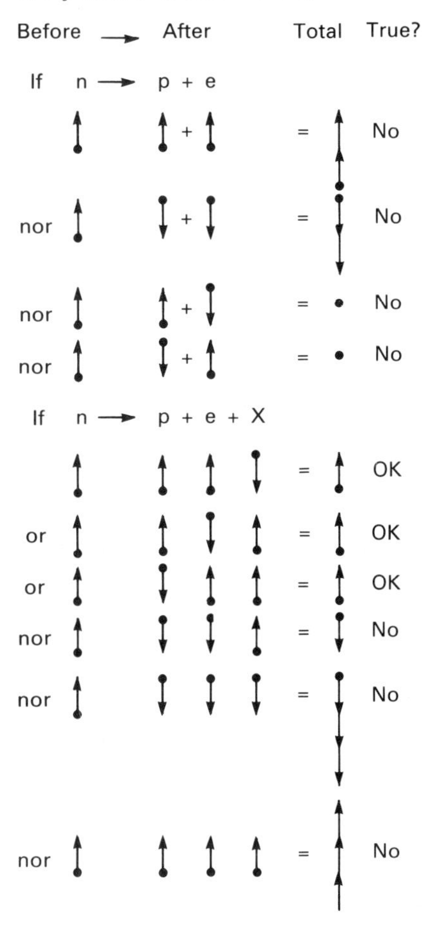

$\frac{1}{2}=\frac{1}{2}-\frac{1}{2}+\frac{1}{2}$ is also true and a valid possibility.

The third combination, in which the proton and electron sum to -1 'down', would not be allowed, because the additional spin $\frac{1}{2}$ could not convert $-\frac{1}{2}$ to $+\frac{1}{2}$.

$\frac{1}{2}=-\frac{1}{2}-\frac{1}{2}+\frac{1}{2}$ is not true,
nor is $\frac{1}{2}=-\frac{1}{2}-\frac{1}{2}-\frac{1}{2}$.

Thus there were some combinations in which spin conservation worked, even if some did not.

Pauli's invention of the new particle satisfied all the physical laws of the conservation of energy, charge and spin.

Neutrinos and their rest mass

One question which was immediately posed was – what is the mass of the new particle? This was answered by examining the energy spectrum of the electrons.

Just as before, energy, including mass–equivalent energy, must be conserved:
rest-mass energy + kinetic energy of neutron = rest-mass energy + kinetic energy of proton + rest-mass energy + kinetic energy of electron + rest-mass energy + kinetic energy of X.

And, just as before, the neutron and proton were always stationary and had no kinetic energy.

So the energy equation in effect read:
rest-mass of neutron = rest-mass energy of proton + rest-mass energy + kinetic energy of electron + rest-mass energy + kinetic energy of X.

When the electrons had their maximum value of kinetic energy, the new particle had its minimum value of kinetic energy, i.e. zero:

rest-mass energy of neutron = rest-mass energy of proton + rest-mass energy + maximum kinetic energy of electron + rest-mass energy of X.
Rearranging the last equation:
rest-mass energy of neutron + rest-mass energy of proton + rest-mass energy + maximum kinetic energy of electron = rest-mass energy of X.

In words, the energy gap between, on the one hand, the energy of the neutron, and, on the other hand, the rest energies of proton and electron, plus the electron's maximum kinetic energy, was equal to the new particle's rest-mass energy. Since the rest mass is equivalent to an energy, the mass of particles is often expressed as the energy equivalent, rather than, say, as grams. The common unit for the rest mass of fundamental particles is the electron-volt (eV).

To everyone's surprise the answer was, as near as could be measured, that its mass was zero, certainly small.

Thus the particle, being small and neutral, acquired the name *neutrino*, this being a construction by the Italian physicist Enrico Fermi and meaning 'little neutral thing' (-ino is an Italian diminutive, as in *bambino*.)

So the equation for beta decay now read:

$$n \rightarrow p + e^- + \text{neutrino, symbol } \nu.$$

The most recent attempts to measure in the laboratory the mass of the neutrino have used virtually the same experiment as the historic one (Sutton 1988). Two recent examples use the radioactive decay of tritium. This is a rare version of the nucleus of hydrogen in which a proton inhabits the nucleus together with two neutrons. Tritium is, like hydrogen, a gas at normal temperatures – large volumes of tritium gas would be needed to contain enough tritium nuclei to view a significant number of decay events. So the experiments which watch how the tritium loses a neutrino in beta decay sometimes lock the tritium conveniently in the molecules of a solid. In an experiment at the Moscow Institute for Theoretical and Experimental Physics, Valentin Lubimov and his coworkers used valine, an organic molecule, $C_5H_{11}NO_2$, in which one of the hydrogen atoms has a tritium nucleus. At the Swiss Institute for Nuclear Research in Zurich, Walter Kundig's group implanted the tritium in carbon. These experiments gave somewhat contradictory results.

One reason for the discrepancies is that if tritium is locked up in solids some energy can be mopped up in exciting the molecules or crystals. The missing energy looks as if it has converted to rest mass of the neutrino. John Wilkerson's group at New Mexico's Los Alamos National Laboratory took the obvious way out and cooled tritium gas (T_2, like H_2) to 160 K in order to decrease the volume of tritium to manageable proportions and to increase the number of decay events which they measured. The decays did excite the tritium molecules but this was allowed for in the calculations. Even so, the upper limit to the mass of the neutrino was still 27 eV. The experiments are summarised in Table 7.

It has to be remembered that these measurements of the mass of the neutrino refer to the neutrinos which are emitted through beta decay. There are other

Table 7. *Rest mass of neutrinos*

Date	Scientist	City	Mass (eV)	Rest-mass scale 0 . . . 10 . . . 20 . . . 30 . . . 40 . . . 50
1980	Lubimov	Moscow	30 ± 16	
1986	Kundig	Zurich	< 18	
1987	Lubimov	Moscow	$30(+2, -8)$	
1987	Wilkerson	Los Alamos	< 27	

kinds of neutrinos, as we shall see in the next section, and their masses are much less certainly known.

Neutrinos in the 1980s

The modern picture of neutrinos has developed somewhat since their 'invention', as an arbitrary experimental requirement, 50 years ago. There are now thought to be at least three and perhaps six kinds of neutrinos. If this is more complicated than just one kind, at least the six kinds fit within a coherent picture, or 'model' of matter (Close, Marten and Sutton 1987).

First to be added to the neutrinos invented by Pauli were the anti-neutrinos. The neutrinos invented by Pauli explained the properties of beta decay:

$$n \rightarrow p + e^- + \nu.$$

Different sorts of neutrinos, whose effects were first found in a laboratory in 1956, participated in a reaction called inverse beta decay in which a proton is induced by a neutrino to form a neutron and a positron:

$$p + \bar{\nu} \rightarrow n + e^+.$$

These neutrinos were called anti-neutrinos. Anti-particles have both equal and opposite properties to the particles: equal masses and spins, opposite charges and magnetic moments. Since the charge on neutrinos is zero, there is no difference between neutrinos and anti-neutrinos in this respect. Moreover, the magnetic moment of the neutrino so small as to be immeasurable, so that neutrinos and anti-neutrinos cannot in practice be distinguished in this respect either. It follows that there are actually no known experimental differences between neutrinos and anti-neutrinos and they might be the same thing, just as photons and anti-photons are the same. But the theoretical formalities are preserved until the situation becomes clear and physicists continue to talk of neutrinos and anti-neutrinos.

Table 8. *Leptons*

Name	Symbol	Mass	Charge	Spin
electron	e^-	0.51 MeV	−1	$\frac{1}{2}$
positron (anti-electron)	e^+	0.51 MeV	+1	$\frac{1}{2}$
muon	μ^-	105 MeV	−1	$\frac{1}{2}$
anti-muon	μ^+	105 MeV	+1	$\frac{1}{2}$
tauon	τ^-	1.8 GeV	−1	$\frac{1}{2}$
anti-tauon	τ^+	1.8 GeV	+1	$\frac{1}{2}$
electron neutrino	ν_e	0	0	$\frac{1}{2}$
electron anti-neutrino	$\bar{\nu}_e$	0	0	$\frac{1}{2}$
muon neutrino	ν_μ	0	0	$\frac{1}{2}$
muon anti-neutrino	$\bar{\nu}_\mu$	0	0	$\frac{1}{2}$
tauon neutrino	ν_τ	0	0	$\frac{1}{2}$
tauon anti-neutrino	$\bar{\nu}_\tau$	0	0	$\frac{1}{2}$

This doubled the number of neutrinos from one to two. The next stages of the development of particle physics theory added four more.

According to the so-called 'standard' model of particle physics, matter is built up from fundamental particles called quarks and leptons. Protons and neutrons for example are clusters of three quarks. Quarks and leptons are held together in clusters by four kinds of fundamental forces: in order of increasing strength the forces are gravity, electromagnetic force, the so-called weak force and the so-called strong force.

The forces are carried ('mediated') by other particles called gauge bosons: they are photons, gluons, gravitons and the W- and Z-particles.

Leptons include the particles of Table 8, and the gauge bosons the particles of Table 9.

The electromagnetic force is the force between moving charged particles, such as electrons moving in orbit around nuclei in atoms; the negatively charged electrons and the positively charged protons attract one another by the exchange of photons. The photons are said to 'mediate' the electromagnetic force.

The strong force is the force which governs what happens inside a nucleus; the protons and neutrons are 'glued' together in a nucleus with gluons.

The graviton, which is supposed to carry the force of gravity, is conjectural – no-one knows if it really exists.

The weak force is the one which is responsible for radioactivity, and it is

Table 9. *Gauge bosons and the four forces they mediate*

Name	Force	Symbol	Mass	Charge	Spin
photon	electromagnetic	γ	0	0	1
W^+	weak – charged current	W^+	83 GeV	+1	1
W^-	weak – charged current	W^-	83 GeV	+1	1
Z	weak – neutral current	Z	93 GeV	0	1
gluon	strong	g	0	0	1
graviton	gravity	not yet shown to exist			

mediated by three kinds of particles – two of them are called W-particles and the third is called the Z-particle. The W-particles are charged, and the force which they mediate produces the 'charged current interaction'. The Z-particle is neutral and its force produces the 'neutral current interaction'.

The neutrinos 'invented' by Pauli to explain the properties of beta decay were accompanied by the emission of an electron; in Table 8, these neutrinos are called 'electron neutrinos':

$$n \rightarrow p + e^- + \nu_e$$

The anti-neutrinos which participated in inverse beta decay were accompanied by the emission of a positron:

$$p + \bar{\nu}_e \rightarrow n + e^+$$

It is because these two kinds of neutrinos are associated with an electron or anti-electron (positron) that they are called electron neutrinos and electron anti-neutrinos. But in the 1950s it became clear that there were other sorts of neutrinos associated with similar reactions involving muons: they became known as muon neutrinos and muon anti-neutrinos, and were observed in 1962. In 1975 a new lepton, the tauon, was discovered and physicists confidently expect that it has its own neutrinos associated with it, the tauon neutrinos; but tauon neutrinos and tauon anti-neutrinos have not yet been observed.

There may be other leptons which have not been discovered and they would presumably have neutrinos associated with them.

There are therefore three kinds or 'flavours' of neutrinos associated with electrons, muons and tauons respectively; each of the three kinds of neutrino

Table 10. *Neutrino flavours*

	Neutrino	Anti-neutrino
Electron	ν_e	$\bar{\nu}_e$
Muon	ν_μ	ν_μ
Tauon	ν_τ	$\bar{\nu}_\tau$

appears in two forms, the neutrino and anti-neutrino, assuming that neutrinos and anti-neutrinos are different. This makes six kinds of neutrinos, and perhaps more if some leptons are undiscovered (Table 10).

Detection of neutrinos

In scientific terms the probability of an interaction between a neutrino and another material particle like a proton is described by a quantity called the cross-section, which represents the effective area for the interaction. The cross-section can be visualised in the following way: if the neutrino passes within this area centred on the proton it will interact with the proton; otherwise not. The cross-section of a neutrino for its interaction with a proton is 10^{-41} cm^2 – this is astonishingly small, but we need to get it into context to see how small it is.

Suppose we wanted to catch neutrinos in a big experimental apparatus full of protons. Water (H_2O) is cheap and made of molecules which include two free protons (the hydrogen nuclei). We could choose water for the detecting material. Suppose, then, a neutrino enters a tank of water.

The molecular weight of water is 18, and its density is 1 g/cm^3, so a mole (18 g) of water occupies 18 cm^3. Avogadro's number is 6×10^{23} so this number of water molecules occupies 18 cm^3. The density of free protons in water is thus $2 \times 6 \times 10^{23}/18 = 7 \times 10^{22}$ protons per cubic centimetre. The total cross-section of the protons in this cubic centimetre is $7 \times 10^{22} \times 10^{-41} = 7 \times 10^{-17}$ cm^2, although the cube itself subtends 1 cm^2 in cross-section. If cubes were placed one behind the other in a long column the cross-sections would gradually accumulate to approach the actual 1 cm^2 cross-section of the water column; at this stage it would be pretty likely that the neutrino would be captured. It would not be absolutely certain, since the neutrino might miss all the protons by chance – on the other hand, the neutrino might be caught by the first proton in the column. If the column accumulated to a length such that the total cross-section for interaction of the neutrino with all the protons accumulated to the actual cross-section of the column it would, however, represent the average path length for the capture of the neutrino. It would take more than 10^{16} cubes in a column to get to this stage. The column would be 10^{16} cm long – this is about 1 light-year. The mean free path for a typical neutrino in water is therefore of the order 0.1 light-year.

If a tank of water could be 0.1 light-year thick, the neutrinos would mostly be soaked up by it. Tanks of water of this size pose severe practical constructional difficulties! Detection of neutrinos is therefore not easy. Neutrinos reach the parts which other particles cannot. The materials of most detectors are virtually transparent to neutrinos, which pass through them, almost never interacting.

First detection of terrestrial neutrinos

Nonetheless neutrinos have been detected: the first detection came in 1956 some 26 years after Pauli first predicted their existence. Los Alamos scientists Clyde Cowan and Fred Reines set up what they mockingly named Project Poltergeist (Theoretical Division) because of the difficulty of detecting their elusive, theoretical quarry. They constructed tanks of water (or its proton-rich equivalent, 'liquid scintillator') near the Savannah River nuclear reactor in South Carolina. The reactor was an abundant source of electron anti-neutrinos. These could interact with the protons in hydrogen atoms in the inverse beta decay reaction:

$$p + \bar{\nu}_e \rightarrow n + e^+.$$

The positron would quickly encounter an electron, when, under terrestrial conditions, it would annihilate into two gamma-rays:

$$e^+ + e^- \rightarrow \gamma + \gamma.$$

The neutron would wander slowly from the tank containing the protons, bumping into other nuclei and slowing down; Cowan and Reines sandwiched the tank of water in a solution of cadmium chloride, to provide a supply of cadmium nuclei which could mop up the neutron, after it had slowed. When the neutron was absorbed by a cadmium nucleus it fired off gamma-rays too. The interval between the two bursts of gamma-rays was calculated to be 5 μs (millionths of a second). Double bursts of gamma-rays with this interval between would prove that the electron anti-neutrino existed (and presumably the electron neutrino too).

Neutrino astronomy

Experiments like these did indeed prove that neutrinos existed and were made in terrestrial sources like nuclear reactors. Neutrinos are also made in

Table 11. *The proton–proton chain of hydrogen burning in the Sun*[a]

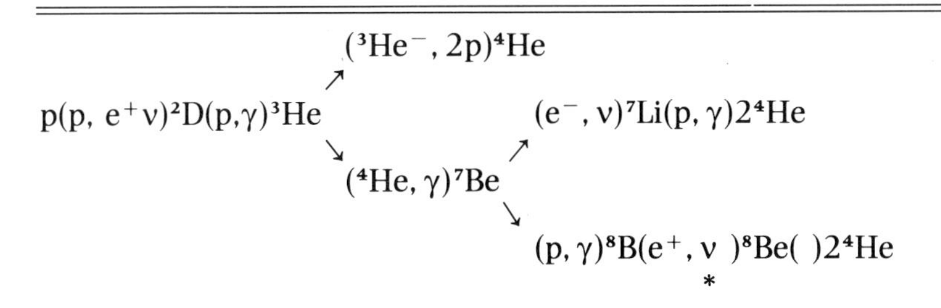

[a]The notation A(b,c)D(,ef)G means $A+b \rightarrow D+c$ and that subsequently $D \rightarrow G+e+f$, with obvious variations of the same idea.

extraterrestrial sources and detecting them is the subject of observational neutrino astronomy.

Observational neutrino astronomy is so far the study of two kinds of objects, represented by one example each:

(i) stars, i.e. the Sun,

and, as of 1987 February 23,

(ii) supernovae, i.e. SN 1987A.

Solar neutrinos are created in the nuclear reactions in the interior of the Sun. Mostly they are low energy neutrinos, but some are of high enough energies to detect. Unlike massive stars (Chapter 5) which burn hydrogen by the CNO cycle, the Sun burns hydrogen by the proton–proton chain, which progressively builds up helium nuclei from protons. The neutrinos come in the main from the decay of boron produced as a by-product for the proton–proton chain. The chain is given in Table 11.

The neutrino marked with an asterisk in Table 11 has 7.2 MeV of energy and is produced by decay of the boron (^8B) nucleus. As well as detection in water, it can be detected by its interaction with the chlorine nuclei in tetrachlorethylene C_2Cl_4 (dry cleaning fluid). The ^{37}Cl nuclei form ^{37}A. The argon nucleus can be flushed from the tetrachlorethylene and detected by its radioactive decay.

An amazing experiment centred around 380 000 litres of tetrachlorethylene in the Homestake gold mine in South Dakota has been run by Raymond Davis since 1970. Neutrino physics suggests that 3.6–8.0 Solar Neutrino Units (SNU) should be detected. The number detected has been in reality between 1.2 and 3.0 SNU (Bahcall, Cleveland, Davis and Rowley 1985). The discrepancy is not good news, since it means that something is wrong with the calculations of neutrino or solar physics, and neutrino astronomy's first test highlighted astronomers' ignorance.

The discrepancy is being confirmed by the Kamioka Neutrino Observatory described in Chapter 8 (Kajita 1987).

Astronomers' self-confidence fared better than they could have hoped in the second detection in neutrino astronomy – the LMC supernova.

Neutrino creation in core collapse – hotter than the inferno

Neutrinos are produced in supernovae in enormous numbers. They are produced in two ways (Shapiro and Teukolsky 1983).

Neutrinos are produced by neutronisation, released when electrons are captured by any free protons in the core and make neutrons:

$$p + e^{-} \rightarrow n + \nu_e.$$

These neutrinos are electron neutrinos. The protons and electrons are forced together during the collapse of the core of the supernova progenitor. Moreover, as the core gets smaller, the energy of the electrons in the core increases and they interact more and more frequently with the protons. The protons and electrons are brought together within the range of the weak interaction. They readily combine in a cluster to make a neutron.

Most of the protons in the core are bound into nuclei – principally iron nuclei. We can describe a nucleus containing Z protons and $A-Z$ neutrons (A nucleons) by the pair of numbers (Z, A). Iron nuclei have 56 nucleons of which 28 are protons, so they would be represented by (28, 56). If the nucleus captures an electron, a proton disappears, and a neutron appears, the number of nucleons remaining the same. The new nucleus is thus ($Z-1$, A). The reaction is:

$$e^{-} + (Z, A) \rightarrow (Z-1, A) + \nu_e.$$

Neutrinos are also produced by supernovae thermally – in the high temperatures in the collapsing material ($T = 10^{10}$ K). We are told in the Bible that Hell is the temperature of melting brimstone (a couple of thousand degrees): according to Kirshner (1988a), tongue in cheek, this implies that supernovae are 10 million times hotter than Hell.

Few people are authoritatively able to describe Hell, nor yet the cores of supernovae. But the cores of supernovae are more amenable to calculation. We are all accustomed to being able to see moderately hot objects – fires, electric light bulbs, the Sun, stars. The temperatures of the surfaces of such objects are in the range 500–10 000 K (more or less). They cool and lose energy by radiating photons – light. Cooler objects (100 K) radiate infrared photons, which we cannot see but which we can sense as they are absorbed by our skin. Even cooler objects (10 K) radiate microwave radio emission. At the opposite extreme, objects whose temperatures are over 100 000 K radiate ultraviolet radiation, X-rays (1 000 000 K) or gamma rays (hotter still). High energy objects in astronomy which radiate X- and gamma-rays include neutron stars and black-holes.

Rare objects, like the collapsing cores of supernovae, have densities measured in millions of tonnes per cubic centimetre and temperatures of 50 000 million degrees Kelvin which exceed temperatures of mere millions of degrees by such amazingly large factors that scientists have little experimental or observational knowledge of them. The astrophysicists rely on theoretical calculations of how such very hot, dense objects cool. One of the reasons why the supernova in the LMC had such an impact on physicists is that it enabled them to flex their intellectual muscles on this problem, testing their understanding of processes, which they had calculated without firm knowledge. Because they turned out mostly right, the physicists gained confidence in themselves and in their science.

The collapsing core of a supernova cools by radiating neutrinos, rather than photons. In the following processes, energy is converted to neutrino–anti-neutrino pairs. All flavours of neutrinos are produced (electron neutrinos, muon neutrinos and tauon neutrinos and their anti-particles), all in about equal numbers.

For example, in the supernova core are electrons and their anti-particles, positrons. They annihilate each other. Usually this reaction produces two gamma-rays, but, under these extremes of temperature and density, the electron and positron make a neutrino and anti-neutrino pair instead. This is called pair annihilation:

$$e^+ + e^- \rightarrow \nu + \bar{\nu}.$$

This interaction is via the weak force. In the production of electron neutrinos and anti-neutrinos, the weak force is mediated by charged W-particles and by the neutral Z^0-particle. Richard Feynman, the Nobel prize-winning physicist, created a method of making sketch diagrams which show how this happens (Figure 24). In the production of muon and tauon neutrinos and anti-neutrinos, only the neutral current interaction works, through the neutral Z^0-particle. It has only been possible accurately to calculate the mechanisms for this since the so-called neutral current interaction was discovered at CERN in 1973.

Other similar sorts of processes which convert energy into neutrino–anti-neutrino pairs include plasmon decay in which quanta of electromagnetic energy propagating in the hot dense plasma are the source of energy:

$$\text{plasmon} \rightarrow \nu + \bar{\nu}.$$

A photon may also convert to a neutrino–anti-neutrino pair in the vicinity of another particle like an electron. This is called photoannihilation, and leaves a degraded gamma-ray (of less energy than originally) as well as two neutrinos:

$$e^- + \gamma \rightarrow e^- + \gamma + \nu + \bar{\nu}.$$

A high speed energetic electron may also collide with a nucleon and transfer some of its kinetic energy into a neutrino–anti-neutrino pair. The electron survives the collision but has lost some of its energy, so this process is known by a German name meaning 'braking radiation', or bremsstrahlung:

$$e^- + \text{nucleon} \rightarrow e^- + \text{nucleon} + \nu + \bar{\nu}.$$

Neutrino diffusion

In the core of a supernova itself the density is so high (100 million tonnes per cubic centimetre) that neutrinos are momentarily trapped by processes which take place through weak interactions. The neutrinos bounce off the electrons in the core like billiard balls, again via the neutral current interaction with the Z-particle mediator. Since they sometimes travel back towards the centre of the core this momentarily slows their outwards drift. The neutrinos take a few seconds to diffuse out from the dense core.

They then get to the lower density outer region of the core (where the density is only 100 000 tonnes per cubic centimetre) where they are radiated from the

Fig. 24. Feynman diagram of the production of a neutrino pair from the mutual annihilation of an electron and positron via the intermediate stage in which a Z° particle is created (the neutral current mechanism).

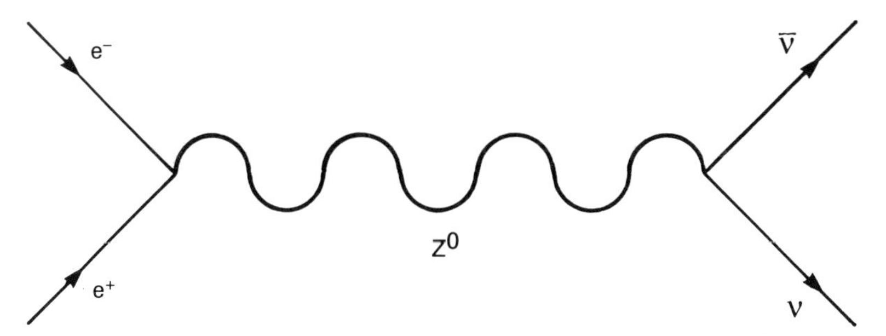

'neutrino-sphere'. This surface is analogous to the solar photosphere, and the whole process is very like the diffusion of energy from the interior of the Sun, before it breaks out of the Sun and is radiated towards us.

In the Sun, photons are created deep in the interior and diffuse outwards and upwards. The centre of the Sun is very hot (tens of millions of degrees) and the photons form an energy distribution appropriate to these temperatures. As they pass into cooler layers of material, the photons cool too.

In the Sun the diffusion time is thousands of years, even though the size of the Sun is only 2 light-seconds or so. The reason why the photons take so long to diffuse this short distance is that they travel back and forth over the same part of the body of the Sun over and over again. They diffuse on until they reach the photosphere.

The photosphere is the surface at which there is a good chance that, rather than interacting again with some of the solar material and being turned back towards the centre of the Sun, a photon will escape from the material of the Sun. If it does, it carries the characteristics (particularly temperature) of that surface, as it travels through the material above the photosphere without interacting much.

When solar astronomers obtain the spectrum of the Sun and determine that the light and infrared photons from the Sun form a distribution which is broadly typical of a black body of temperature 6000 K, they are telling us that this is the temperature of the material at the solar photosphere. They could express this temperature equivalently by telling us the energy of the photons (on average). Expressed in units of electron-volts (eV – the energy of an electron falling down an electric potential of 1 V), sunlight has an energy of 0.5 eV.

When the neutrinos created in a supernova leave the collapsing core at the neutrino-sphere the material nearby has a temperature of 4×10^{10} K. The neutrinos form a thermal distribution with this temperature. The neutrino energies are 3.5 MeV (1 MeV = 1 000 000 eV).

Number

The number of neutrinos produced can be estimated to a gross order of magnitude quite simply. Consider only the neutrinos produced by neutronisation. One neutrino is produced for every proton converted to a neutron. The mass of the collapsing core is about the mass of the Sun, 2×10^{33} g. Avogadro's number is 6×10^{23} so the solar mass contains of the order of 10^{57} protons and therefore releases 10^{57} neutrinos. The contribution of thermal neutrinos to the neutrino generation increases this number by about a factor 10 to a total number for the neutrinos produced of

$$N = 5 \times 10^{58}.$$

The number of neutrinos and their average energy of 3.5 MeV can be used to determine how much energy was carried off from the supernova.

$$E = N \times 3.5\ \text{MeV}$$
$$= 5 \times 10^{58} \times 3.5 \times 10^{6} \times 1.6 \times 10^{-19}\ \text{J}.$$

This implies that 3×10^{46} J (3×10^{53} erg) was carried off by neutrinos from the supernova. If we use the equation $E = mc^2$ (where c = the speed of light), which relates this energy to mass, we find

$$m = E\ \text{erg}/c^2\ (\text{cm/s})^2 = 3 \times 10^{53}/(3 \times 10^{10})^2 = 3 \times 10^{32}\ \text{g}.$$

This energy is thus equivalent to the total annihilation of more than 0.1 solar masses of material (1 solar mass $= 2 \times 10^{33}$ g).

The neutrinos are emitted from the core collapse in just two or three seconds. The neutrino luminosity of the supernova is thus

$$L = 1 \times 10^{53}\ \text{erg/s}.$$

As we saw in Chapter 6, the luminosity of the Universe radiated in light and infrared emission from stars is

$$L = 10^{10}\ \text{galaxies per Universe} \times 10^{11}\ \text{stars per galaxy} \times 4 \times 10^{33}\ \text{erg/s per star} = 4 \times 10^{54}\ \text{erg/s}.$$

In table 3 in Chapter 5 it was estimated that there is one Type II supernova per 100 years per galaxy of 10^{10} solar luminosities. The neutrino luminosity of the Universe is thus

$$L_\nu = 10^{10}\ \text{galaxies per Universe} \times 10\ \text{SN per galaxy of}\ 10^{11}\ \text{solar luminosities per century} \times 1 \times 10^{53}\ \text{erg/s of neutrino luminosity per supernova}.$$

The number of seconds in a year is approximately† $\pi \times 10^7$. So

$$L_\nu = 1 \times 10^{64} \times 1/(100 \times \pi \times 10^{7})\ \text{erg/s}$$
$$= 3 \times 10^{54}\ \text{erg/s}.$$

† The number of seconds in a year is

365.25 days per year × 24 hours per day × 60 minutes per hour × 60 seconds per minute = 31 557 600.

This is approximately $\pi \times 10^7$, by coincidence. A good test to see whether a class is awake is to show this calculation and to remark that the number of seconds in a year is related to π only approximately because the Earth's orbit around the Sun is eccentric. A frown of puzzlement at this absurd explanation will slowly grow deeper on the faces of those who are thinking.

The neutrino luminosity of the Universe from supernovae is thus about the luminosity radiated in light and infrared radiation from stars.

Number of neutrinos received from the supernova at Earth

The flux of neutrinos at the Earth can be calculated assuming that they spread uniformly over the surface of a sphere whose centre is the LMC and whose radius is the distance from the LMC to the Earth, 170 000 light-years. Since 1 light-year $= 9.5 \times 10^{15}$ m, the surface area of this sphere is

$$A = 4 \times 3.14 \times (170\,000 \times 9.5 \times 10^{15})^2 \text{ m}^2$$
$$= 3.2 \times 10^{43} \text{ m}^2.$$

Fig. 25. The neutrinos from the LMC travelled outwards on the surface of a sphere which had a radius $R = 170\,000$ light-years at the moment that some neutrinos passed through the Earth.

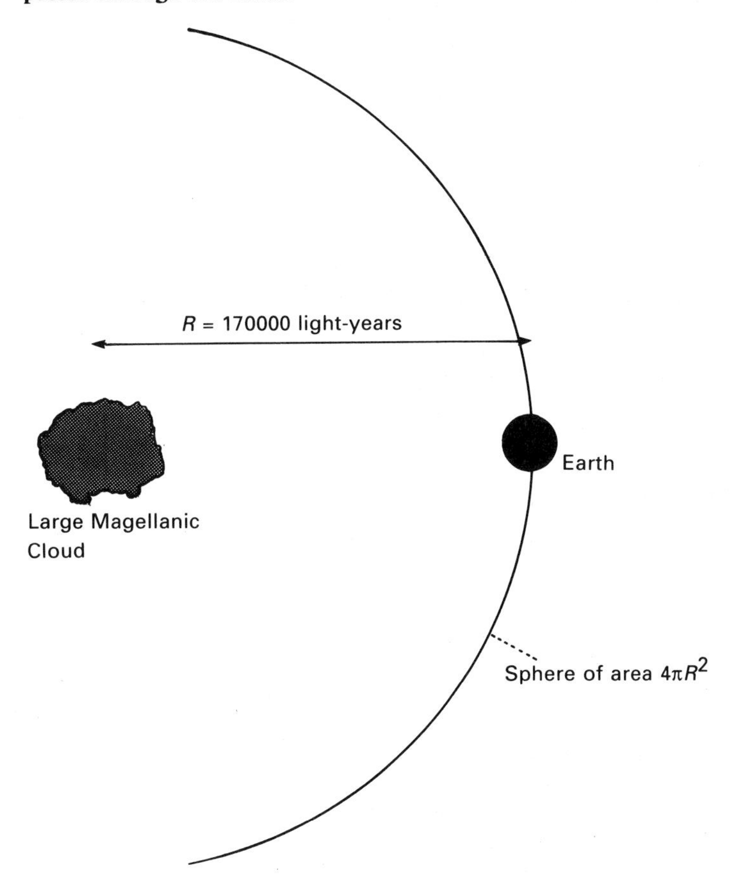

The flux was therefore

$$n = 5 \times 10^{58}/A \text{ neutrinos per square metre}$$
$$= 10^{15}.$$

The flux of neutrinos from the LMC supernova was, thus, 100 thousand million per square centimetre at Earth.

Bearing in mind that, depending on whether people are viewed from above or face-on or sideways, and depending whether they are fat or thin, tall or short, nevertheless most people are of order 1 m^2 in cross-section, the core collapse of a star in another galaxy from ours irradiated everybody on this planet with about 100 million million neutrinos on 1987 February 23.

The average energy of the neutrinos was 5 MeV – this is the same as 3 million electrons passing along a wire connecting the poles of a 1.5 V battery cell. Thus each neutrino carries a microscopic amount of energy: 8×10^{-13} J – comparable to the energy of a flea stepping over a hair. But the number of neutrinos is so large that the flux of energy carried by the neutrinos was significant. It was:

$$F = 8 \times 10^{-13} n \text{ J/m}^2$$
$$= 8 \times 10^{-13} \times 1 \times 10^{15} = 800 \text{ J/m}^2.$$

The pulse of neutrinos lasted a second or two, and so this energy flux was of order a few 100 J/m^2 – say about half of the solar flux of radiated energy, which is 1400 J/m^2 per second. Worldwide, the energy which passed through the Earth was 3×10^{16} J, comparable to the energy released in a megaton atomic bomb.

In spite of these gigantic numbers of neutrinos and frightening amounts of energy which they carried, the wave of neutrinos swept across the Earth on 1987 February 23 with completely negligible effect on all human affairs except astronomy.

The reason that we didn't feel a thing is that, since the Earth is much smaller in size than the mean free path of the neutrinos in material, the Earth, including we ourselves and our sensory organs, are virtually transparent to neutrinos. Very little energy indeed was lost by the wave of neutrinos and deposited in the Earth during the encounter – although the energy carried through the Earth by the neutrinos was of the order of that of a nuclear bomb, the effect of the energy deposited in the Earth by the neutrinos was as ignorable on a global scale as a car crash.

The impact of the neutrinos on astronomy was not ignorable.

8
The neutrino observatories and the supernova neutrinos

Reports of the detection of 30 or so neutrinos from the LMC supernova have been issued by 83 scientists representing 30 institutions and working at four observatories (Table 12). Reading the scientists' accounts of their work, it seems as if there was a mischievous jinx at work, putting stumbling blocks in the way of detecting these neutrinos. But evidently, if there was a teasing Puck, there was also a Good Fairy who kept the experiments' clocks ticking, if inaccurately; who kept most of the photomultipliers going somehow; and who arranged for calibration tests which interrupted the operation of the detectors to commence minutes after detection of the neutrinos which arrived at Earth after their 170 000 year journey.

Neutrino observatories look nothing like optical observatories: the first difference is their situation. We are accustomed to the idea that astronomical observatories are on the tops of mountains, but neutrino observatories are well below ground. The neutrino events which they could detect would otherwise be lost in a confusing welter of sporadic and random background events caused by cosmic rays showering the Earth's surface from above. In particular cosmic ray muons may interact with the rock near the detector, making gamma-rays and neutrons, which can give spurious signals. So neutrino astronomers put on hard hats and burrow into the dark recesses of the Earth, working alongside miners and tunnellers and, up to 1987 February 23, only dreaming of the faint traces which they might one day glimpse from a distant exploding star, far above the ground.

Table 12. *Supernova neutrinos and their scientists*

Neutrinos	Scientists
1987 February 23 02:52 UT	
5 neutrinos with energies between 7 and 11 MeV were detected within 7 s by the Mt Blanc Underground Neutrino Observatory (Aglietta *et al.* 1987a, b).	20
1987 February 23 07:35 UT	
12 neutrinos with energies between 8 and 35 MeV were detected within 10 s in the detector Kamiokande-II in Japan (Hirata *et al.* 1987).	23
8 neutrinos with energies between 20 and 40 MeV were detected within 6 s by the IMB detector in Ohio (Bionta *et al.* 1987; Svoboda 1987).	36
5 neutrinos with energies between 13 and 18 MeV were detected within 9 s by the Baksan group in the USSR (Alexayev, Alexayeva, Krivosheina and Volchenko 1987; Pomansky 1987).	4

The Mt Blanc Neutrino Observatory

First to announce the detection of neutrinos (Castagnoli 1987) – but the most controversial – was the Underground Neutrino Observatory under Mont Blanc, or Monte Bianco, in a service tunnel for the road tunnel which connects Italy and France. Unlike the other two main observatories, this one was specifically set up in 1984 to detect neutrinos from a supernova – the next supernova in our Galaxy was what the scientists had in mind, not one in the LMC. It is run by the Istituto di Cosmogeofisica and the Istituto di Fisica Generale in Turin and the Institute of Nuclear Research in Moscow (Aglietta *et al.* 1987a).

The massive bulk of Europe's largest mountain, Mt Blanc, above the tunnel, forms a very large screen against cosmic ray events – there is the equivalent of 5200 m of rock above the detector. So most cosmic ray muons interact with the mountain well above the detector and not near to it. In addition there is 200 tonnes of iron shielding which surrounds the detector to protect it against the natural radioactivity of the mountain rock. The detector records a background contamination of only about 3.5 events an hour.

The detector itself is 90 tonnes of liquid scintillator, a collection of organic molecules which are rich in hydrogen atoms, and therefore free protons. The molecules are chains of carbon atoms with two hydrogen atoms attached:

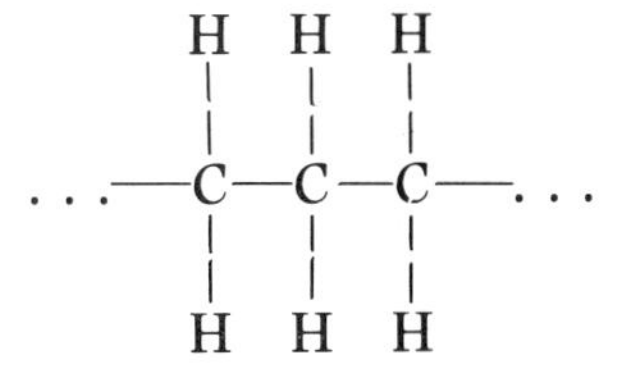

There are two extra hydrogen atoms at the ends of the molecule. The chemical formula of liquid scintillator is thus C_nH_{2n+2} with n about equal to 10 – different molecules may have different numbers of carbon atoms in the chain. The 90 tonnes of liquid scintillator contains 8×10^{30} protons. The 90 tonnes is divided among 72 individual counters arranged in a block which is 7 m wide × 8 m long × 5 m high.

The detector watches for electron anti-neutrinos through the inverse beta decay process, which creates an energetic positron:

$$p + \bar{\nu}_e \rightarrow n + e^+.$$

The positron causes a scintillation of light in the liquid which is detected by arrays of photomultipliers, three of them at the top of each counter.

The neutron collides with a proton to give a deuterium nucleus and a gamma-ray of energy 2.2 MeV:

$$n + p \rightarrow d + \gamma.$$

The gamma-ray also generates a scintillation in the liquid; it is less energetic than the positron-induced scintillation and it follows it about 200 μs later, this being the usual time for the neutron to find a proton to merge with.

If three photomultipliers over one of the counters all see a scintillation, the experimenters count it as an event. If it represents an energy of over 6 MeV, it is regarded probably as a positron generated by a neutrino. If, within 500 μs, a second smaller scintillation is seen, representing an energy of over 0.8 MeV, it is regarded as probably the gamma-ray arising from the neutron generated by the neutrino, and the combination of the two events is such an unusual occurrence that it is virtually certain that a neutrino was detected.

In every 100 events (which are accumulated in an hour and a half or so) there are six or so spurious cosmic-ray muon events: the rest are neutrinos of one sort or another. There is always the problem of distinguishing neutrinos produced by radioactive rocks around the detector, neutrinos produced by the interaction of cosmic rays with the Earth and with the atmosphere above the mountain, neutrinos which come from the Sun, the cosmic background of neutrinos left over from the Big Bang – and the neutrinos from a supernova.

The way this is done is to look for evidence that the neutrinos arrived in a burst. The radioactivity, the Sun and the cosmic background are all constant

Fig. 26. How many times did the Mt Blanc Neutrino Observatory see five or more pulses in short bursts? This is the question answered by the left hand curve of this diagram compiled from two days' data which include the neutrino burst of 1987 February 23 at 02:52 UT. There were two bursts of five pulses which had a duration of up to 40 s, three with a duration up to 50 s, ten with a duration of up to 60 s . . . 100 bursts of five pulses with a duration up to 200 s and so on. Other lines and sets of data represent the same question for 10 and for 15 pulses respectively. The theoretical curves, calculated for random pulses, fit the data very well – except for the very first point which stands well above its line. There was thus one burst of five pulses which lasted 7 s and which was very unlikely to be random like all the others. This was the burst of pulses attributed to neutrinos from the supernova.

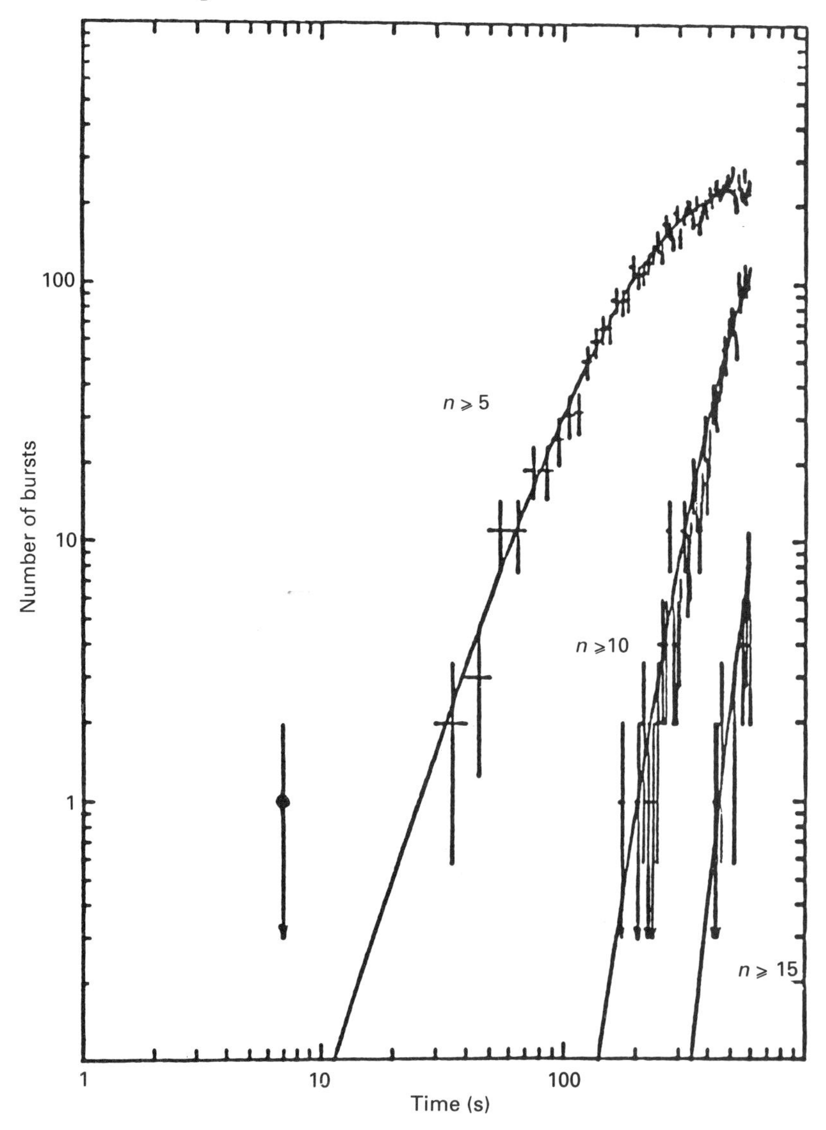

Table 13. *The Mt Blanc neutrinos from SN 1987A*

Event number	Time (1987 February 23 UT)	Energy (MeV)
994	02:52:36.79	6.2
995	40.65	5.8
996	41.01	7.8 + 1 MeV gamma-ray
997	42.70	7.0
998	43.80	6.8

sources of neutrinos. The cosmic rays can arrive in showers but infrequently do so, and mostly appear as a constant background – but supernovae produce neutrinos in a burst of significant numbers lasting a few seconds.

As in all similar particle physics experiments, the evidence is statistical: the experimenters form statistical distributions showing, for instance, how long it takes to see a given number of neutrinos, for example five neutrinos. In the Mt Blanc experiment it usually takes several minutes to see such a number. But less often it is only 1 min. Less often still it is 30 s. Very rarely is it as little as 20 s and almost never is it as little as 10 s. In a run of data gathering it is possible to calculate the probability that a given sequence of neutrino events is due to chance or represents an improbably high number of neutrinos in a short time.

The Mt Blanc experimenters plot and analyse by computer their histograms of probability as the experiment is running. On 1987 February 23 at 02:52:36 UT, the on-line print-out flagged as very improbably due to chance a burst of five neutrinos during 7 s (Table 13). The five occurred in different counters; one of them, the third event, was followed 278 μs afterwards by a low energy scintillation, just as expected from the gamma-ray produced by the neutron in the inverse beta decay interaction.

Having regard just to the number of neutrino events which had been detected within 7 s, the burst at 1987 February 23 02:52 UT could be imitated by the random background of events in the detector once every 1.4 years and was calculated as such by the normal on-line analysis before Shelton's discovery of the supernova. The probability that such an event would occur by chance in the eight hours before the first photograph of the supernova by McNaught is 6.5×10^{-4}. Taking into account all the circumstances of the detection, the sequence was the most unusual combination which the detector had ever seen. The Mt Blanc experimenters (20 scientists from two countries) concluded that they had seen a burst of neutrinos from the supernova in the LMC.

Proton decays and GUTs

In contrast to the Mt Blanc detector, the two detectors known as IMB and Kamiokande are not primarily intended to detect neutrinos from a supernova. They are intended to detect the decay of the proton and contribute to the picture of the Universe held, not by astronomers, but by elementary particle physicists. The detail of the picture which IMB and Kamiokande were intended to answer was whether the proton was immortal or whether it was simply very long lived – the sort of question which might occur to a junior lecturer as he wondered about a professor with tenure.

The decay of the proton is a prediction of a theory of force which combines the electromagnetic, weak and strong forces into a so-called Grand Unified Theory (GUT). Typically the proton might produce a positron and a pion, the latter decaying in turn to two gamma-rays. Looking around us, we have formed the working hypothesis that protons are stable. For example, we do not notice water seething with bubbles of oxygen, liberated as the protons which are the nuclei of hydrogen in the H_2O molecules change to something else. Nor do we die from radiation sickness induced by the decay of the protons in our own bodies. But, according to GUTs, protons are only marginally unstable. The lifetime of the proton is calculated to be something in the vicinity of 10^{30} years; this is why protons are not readily seen to be decaying and, indeed, why there are so many protons still here just 10^{10} years after the Big Bang in which they were formed – not many have had time to decay. Only those few which die young would give evidence that protons do die.

The detectors which have been set up to spot the possible deaths of such long lived objects have to be prepared to capture one death every five years in 700 tonnes of protons (7000 tonnes of water), through a cosmic ray background of 100 million confusing events every year (Svoboda 1987). This is about the same as searching for a needle in a haystack which includes lots of kinds of rubbish – nails, screws, wire, string, sticks etc. (A needle in 100 tonnes of straw is about one object in a billion.)

Some of the events which the detectors see are neutrinos. They come from radioactivity, from cosmic rays, from the Sun and from the Big Bang. The constant background of neutrinos from the Sun was, up to 1987 February 23, one of the most interesting positive detections from the detectors, since so far no proton deaths at all have been seen in any of the detectors set up to watch for one. All of the dozens of events thought initially to be candidates for records of a proton decay have, in the end, turned out to be imitations of proton decays by neutrino events. Thus, proton decays are indeed rare – even rarer than initially predicted.

Indeed, one detector, the IMB, which has been running for several years, has demonstrated that the lifetime of the proton is at least 10^{32} years for the decay to pion and positron, and so has killed one version of a GUT called *SU*(5) which predicted a much shorter time. Another detector, Kamiokande, by detecting no decays of a proton to an anti-neutrino plus a K^+ meson, has proved that this mode has a lifetime of at least 10^{31} years. Thus, the limits which the detectors have generated have clearly advanced particle physics: in addition, the detection of the neutrinos from the SN 1987A has given a new spur to the physicists – so much so, that the Kamiokande project was financed for a design study to construct 'Super-Kamiokande' (10 times bigger than Kamiokande) just six months after the neutrino burst from SN 1987A.

Kamiokande

Kamioka is the name of a zinc mine near Gifu in Japan, 300 km west of Tokyo. Kamiokande (Beier 1986) is an acronym for the *Kamioka* Nucleon Decay Experiment which is installed underground in the mine at a depth equivalent to 2400 m (not so deep as the Mt Blanc detector). It has been operating since 1983 in a first version. Kamiokande-II has been in operation since 1986 (Koshiba 1987).

It is a tank of 3000 tonnes of water (of which 2140 tonnes is effective), surrounded by a set of over 1000 of the world's largest photomultipliers, each 50 cm in diameter, specially developed for the project by Hammamatsu Photonics. The tank is a cylinder 16 m in diameter and 16 m high. The photomultipliers detect scintillations produced by particles entering or produced in the tank. As in the Mt Blanc detector, the location of the detector in the mine is intended to reduce cosmic ray background events produced by mesons; these are one of the three main sources of background, which produce one background event about every 1.5 s. The second source of background is low energy radiation (gamma-rays and neutrons) coming from radioactive decay in the surrounding rocks. The third is radioactivity in salts dissolved in the water itself; the water had to be purified to absorb these radioactive atoms as much as possible.

In the way in which we have now become accustomed, a cosmic electron anti-neutrino may interact with a proton in the hydrogen atoms in the water (H_2O) and produce a neutron and a positron (inverse beta decay):

$$p + \bar{\nu}_e \rightarrow n + e^+.$$

In addition, but with lower probability (about 1% of the probability of inverse beta decay at neutrino energies of 10 MeV), a neutrino of any flavour may

Table 14. *Free protons and electrons in neutrino observatories*

Material	Formula	Protons	Electrons	Ratio
water	H_2O	2	$1\times8+2=10$	5:1
liquid scintillator	$C_{10}H_{22}$	22	$10\times6+22=82$	4:1

scatter on one of the electrons in the water molecule, putting some of its energy into the electron and breaking it free from the molecule:

$$\nu+e^-\rightarrow\nu+e^-.$$

Although any kind of neutrino can do this, electron neutrinos and antineutrinos can do it more effectively than muon or tauon neutrinos and antineutrinos.

Both these reactions can taken place in the liquid scintillator of the Mt Blanc detector. But a simple calculation shows that liquid scintillator has relatively less electrons compared to protons than water, and electron scattering is not quite as important (and there are energy dependencies too) – see Table 14.

Now, the speed of light in water is about $\frac{3}{4}$ the speed of light in a vacuum (referred to as *c*). In each reaction between the neutrino and the water molecules, the electron or positron is given a significant part of the energy of the neutrino, and this causes the electron/positron to travel faster than the speed of light in the water (it travels at a speed somewhere between 0.75*c* and *c* itself). Charged particles, like electrons or positrons, which travel through a medium like water faster than the speed of light in that medium, produce a burst of light called Cerenkov radiation. The Cerenkov radiation caused, in the first instance, by the cosmic neutrinos is sensed by the photomultipliers which surround the tank of water (Plate 5).

Some of the photomultipliers in Kamiokande are within the water looking outwards to view background events (like cosmic rays) which originate from outside; these can then be ignored. The active mass of water in which cosmic events are certainly seen is called the fiducial mass of Kamiokande and is 800 tonnes.

The Kamiokande-II detector notices electrons/positrons created by neutrinos of energy over 5 MeV, but the background events are confusing unless the energies are abovc this level, say 8 MeV or more.

The Cerenkov radiation forms a burst of light which is radiated in a hollow cone whose axis lies in the direction of motion of the electron or positron. The angle of the cone is about 42°, and for every centimetre of travel through the water the electron/positron emits some 200 light photons. The photomultipliers can distinguish Cerenkov radiation from background events by the elliptical pattern of light from a section across the hollow cone. From the pattern, and how elliptical it is, the axis of the cone of the Cerenkov radiation can be determined, to an accuracy typically of 20°, so that the direction of flight of the electron or positron can be determined (Plate 5).

Thus Kamiokande is a neutrino telescope which can point in a direction. This is useful particularly for the electrons produced by electron–neutrino scattering.

Electrons are light and loosely bound; when one is hit by a neutrino it 'remembers' the neutrino's momentum, just as a small jack is pushed directly forward when hit by a massive ball in the game of bowls. This means that the

Fig. 27. The detector Kamiokande-II, with dimensions in mm, is a steel cylindrical tank containing water. Water also acts as an anti-counter (area with dashed lines). Photomultiplier tubes line the inner and outer walls, looking inwards to see Cerenkov radiation from inside the tank and outwards to guard against charged particles entering from outside.

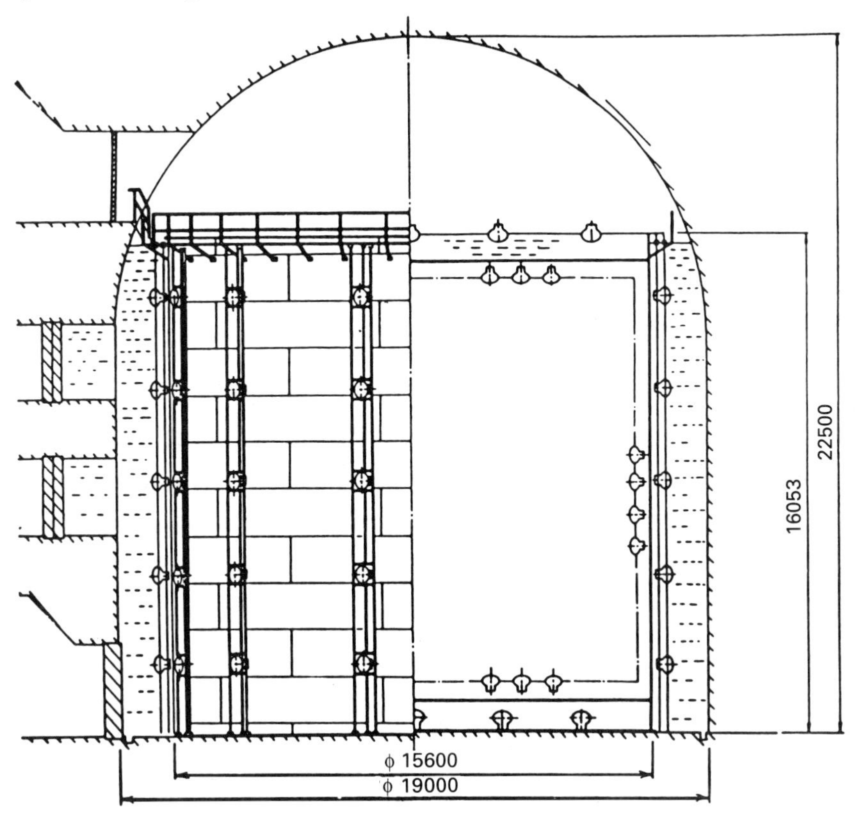

Cerenkov radiation from electron scattering is beamed forward in the direction of the incoming neutrino. This in turn implies that Cerenkov radiation from an electron scattered by a neutrino from SN 1987A points away from the LMC, which at the time was 20° below the horizon at Gifu, Japan, where the Kamioka mine is.

On the other hand, since protons are very heavy, positrons created by interactions between the electron anti-neutrinos and protons are isotropic, just as a small jack which strikes a heavy bowling ball ricochets at a wild angle.

When the Kamioka group (23 scientists at six institutions) heard of Shelton's discovery of the supernova they searched their data for neutrinos from SN 1987A (Hirata *et al.* 1987). Kamiokande-II saw 12 Cerenkov events

Fig. 28. Kamiokande and the IMB neutrino observatories are tanks of water surrounded by photomultipliers. Some neutrinos which enter the water produce relativistic electrons (and positrons) which generate Cerenkov radiation. The Cerenkov radiation is produced within a hollow cone, and fires some photomultipliers on the wall of the tank, in an ellipse cutting across the cone. From the geometry of the ellipse, the direction of the relativistic electron can be determined: this might be related to the incoming direction of the neutrino, if the electron 'remembers' the neutrino's momentum well enough.

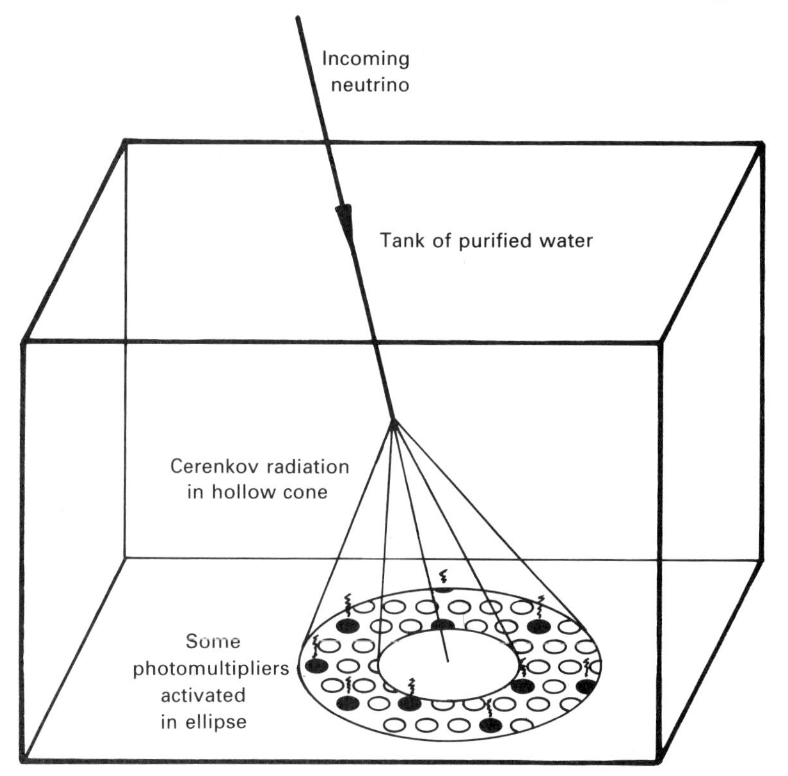

Table 15. *Kamiokande's neutrinos from SN 1987A (1987 February 23 07:35:35 UT)*

Event	Arrival time relative to the first (s)	Number of photomultiplier tubes hit	Electron energy (MeV)	Angle (°)
1	0.000	58	20.0	18
2	0.107	36	13.5	15
3	0.303	25	7.5	108
4	0.324	26	9.2	70
5	0.507	39	12.8	135
6	0.686	16	6.3	68
7	1.541	83	35.4	32
8	1.728	54	21.0	30
9	1.915	51	19.8	38
10	9.219	21	8.6	122
11	10.433	37	13.0	49
12	12.439	24	8.9	91

representing neutrinos in a burst at 07:35 UT on 1987 February 23 (Table 15). They were actually very lucky to see anything since two minutes later the device turned off for a routine calibration sequence (Bates 1988)!

The precise time that Kamiokande saw the neutrinos is uncertain by up to a minute, since the clock in Kamiokande's computer had been manually set while the operator listened to the time on the telephone – it had seemed unimportant to the Kamioka scientists to record the exact time at which a proton decayed in the tank, since they expected to see it happen rarely – once every few years! Moreover, because of a complete power failure after the detections, which caused the clock to stop, the error in the clock could never be determined (Helfand 1987).

The angle which is tabulated in Table 15 is the angle between the flight path of the electron/positron and the direction from the LMC. Event number 6 was of lowest energy and may have been a background event rather than a supernova neutrino. Nearly all the Cerenkov bursts showed large angles relative to the line of sight from the LMC. The electrons/positrons which produced the bursts were therefore nearly isotropically distributed and were produced by inverse beta decay (and therefore most were positrons) and the neutrinos which were responsible for most events were electron anti-neutrinos. However, within the accuracy of the measurement, events number 1 and 2 were from electrons which pointed back to the LMC. They may have done so by chance, but they may also have genuinely remembered the direction of the neutrino which produced them. They may therefore have been from electron scattering, rather than inverse beta decay. They may therefore represent some flavour of neutrino other than the electron anti-neutrino.

The IMB neutrinos

IMB stands for Irvine–Michigan–Brookhaven, which are the homes of the principal institutions which run the IMB detector. It is located under the shore of Lake Erie in an Ohio salt mine at the equivalent depth of 1570 m.

IMB is like a large Kamiokande. It is a tank of 7000 tonnes of purified water, surrounded by 2048 8 in photomultipliers which look for Cerenkov radiation produced in the water by electrons/positrons knocked by cosmic neutrinos, just as the Kamiokande detector does (Svoboda 1987; Bionta *et al.* 1987). It too has guard photomultipliers which look outwards and sense background events entering from outside; its active volume of water is 5000 tonnes. It notices higher energy neutrinos than Kamiokande, with energies above 20 MeV.

Fig. 29. The supernova signal as it was first recorded by Kamiokande in a plot of the number of photomultipliers hit by pulses of Cerenkov radiation. Background events caused 10–20 photomultipliers to fire in a given pulse; the events at 07:35 UT (Time = 0) are clearly very significant and non-random. Because they fired a large number of photomultipliers the Cerenkov electrons are clearly much more energetic than usual for typical Kamiokande background events.

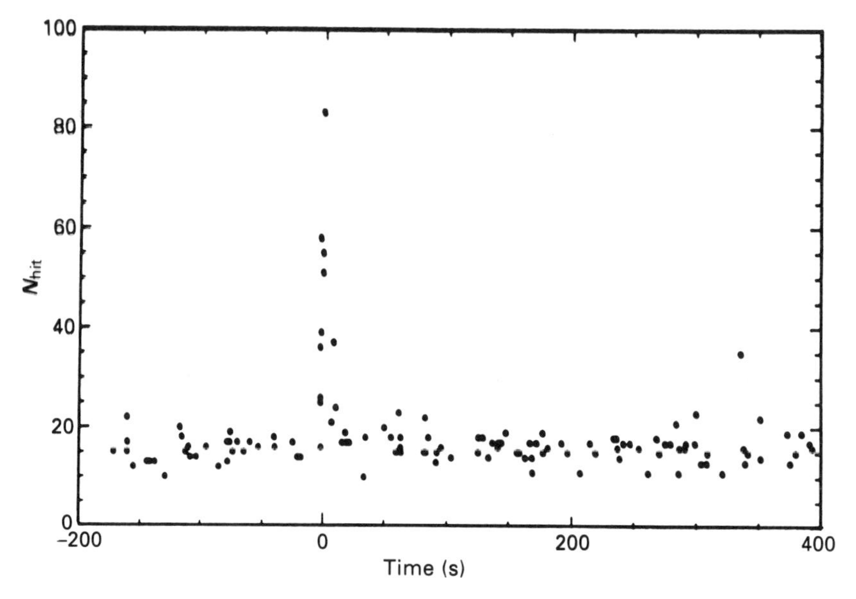

When the IMB scientists learned of the discovery of SN 1987A on 1987 February 24, and the detection of neutrinos from SN 1987A at 02:52 UT on 1987 February 23 by the Mt Blanc Underground Neutrino Observatory, they searched their data tapes for February 21, 22 and 23 for a neutrino burst (van der Velde 1988). The search algorithm screened out unlikely events, and chose only those within the central 3300 tonnes of the detector.

Only four high energy events were found in this quick look, and these were attributed to atmospheric contamination. The IMB scientists found no neutrino events at 02:52 UT, and attributed this to the fact that the Mt Blanc detector had found low energy neutrinos, none of which would be above the IMB energy threshold. The physicists were somewhat discouraged by this; they were further discouraged from looking at later times by the fact that, an hour after the Mt Blanc Underground Neutrino Observatory detected its neutrinos, IMB suffered a partial equipment failure. One of the four power supplies for the photomultipliers failed. The supply was dead for seven hours; so too, therefore, were a quarter of the photomultipliers – 512 photomultipliers in the south wall of the tank and the south-east corner of the top. The remaining photomultipliers continued working properly.

The on-line data-reduction computer programme which monitors the IMB data for interesting events was not able automatically to make an allowance for the failure, and special programmes to process the data had to be written. No-one wanted to do this – what was the point of examining data taken long after the event which had been discovered? – and this task was postponed.

But then the rumour circulated that Kamiokande had found its neutrinos, not at 02:52 but at 07:35 UT. The IMB group (36 scientists at 16 institutions) immediately set to work to modify the computer's data-reduction programme to take account, at this critical time, of the missing photomultipliers. The scientists searched the data beyond the equipment failure – and found nine events at the time of the Kamioka burst. Just one of them was a cosmic ray muon – the rest really were cosmic neutrinos.

So eight neutrinos from SN 1987A had been detected by IMB within 6 s at 07:35 UT (Table 16 – the missing serial numbers in the list represent events which are spurious). The computer had saved the detections but not brought them to the scientists' attention. Its programme was not intelligent enough to take account of the partial equipment failure. The effect of the missing photomultipliers was to reduce somewhat the detection efficiency of the IMB detector, but not to affect the quality of the data much.

Never before (since IMB began to operate in 1982) had even two neutrinos been recorded so close together; the probability that eight would occur by

Table 16. *IMB's neutrinos from SN 1987A (1987 February 23 07:35:40 UT)*

Event no.	Time (UT)	Number of photomultiplier tubes	Energy (MeV)	Angle (°)
33162	07:35:41.37	47	38	74
33164	41.79	61	37	52
33167	42.02	49	40	56
33168	42.52	60	35	63
33170	42.94	52	29	40
33173	44.06	61	37	52
33179	46.38	44	20	39
33184	46.96	45	24	102

chance within 6 s was calculated (Svoboda 1987) to be 10^{-33}, 'a reasonably good approximation to "never".'

The IMB neutrinos in Table 16 are all (except the last) pointing away from the LMC (the angles are all except one less than 90°). Even when the effect of the dead photomultipliers is taken into account, they are not isotropic. But it is difficult to assess which individual neutrinos are electron anti-neutrinos and which may be another flavour.

The Baksan Underground Scintillation Telescope

Since 1980 a Soviet experiment much like the Mt Blanc experiment has been operating below Mt Elbrus in the mountains of the North Caucasus, at a depth equivalent to 850 m, and designed to detect supernovae (Alexayev *et al.* 1987). It too consists of liquid scintillator, 200 tonnes, and detects neutrinos over 10 MeV in energy. It has been live for 5.7 years since its inception but detected no undoubted neutrino bursts from supernovae, not even on 1987 February 23. However, when the scientists running the Baksan Underground Scintillation Telescope heard of the neutrino detections by Mt Blanc, Kamioka and IMB they searched their data for confirmation. They discovered (Table 17) one event close to the time of the Mt Blanc burst and five events 30 s after the time of the IMB burst. These events were not so much above the background that they could be thought significant on their own, but, in spite of the 30 s discrepancy, they may be significant when taken in association with the other detectors.

Table 17. *The Baksan neutrinos from SN1987A (1987 February 23)*

Time of event (UT)	Energy (MeV)
02:52:36	11
07:36:11.8	12
12.3	18
13.5	23
19.5	17
20.9	20

Seeing neutrinos

Could we have seen the supernova's neutrinos?

Of course, human eyes are not sensitive to neutrinos as such. However, eyes contain a vitreous humour, which is essentially water. Electrons or positrons created by neutrinos from the supernova, or indeed anywhere, can create Cerenkov radiation in this water as well as in the water-based neutrino observatories. They radiate a couple of hundred light photons within the eyeball, compared to the threshold of sensitivity of the human eye, which is a few photons. A simple calculation could estimate whether enough electrons or positrons were created by the supernova within the eyeballs of the human race to be detected.

There are 5×10^9 people on the Earth, each with two eyeballs which contain of the order 10 g of vitreous material. Thus, there are 100 000 tonnes of active human eyeballs, worldwide, compared to just 5000 tonnes of water in the IMB detector. If we scale in proportion, then we find that, since eight neutrinos were detected in IMB, the number of neutrinos which produced bursts of Cerenkov radiation within the mass of active human eyeballs was

$$N = \frac{8 \text{ neutrinos} \times 5 \times 10^9 \text{ people} \times 2 \text{ eyes per person} \times 10 \text{ g per eyeball}}{5000 \text{ tonnes} \times 10^6 \text{ g per tonne}}$$

$$= 160$$

We should take into account the fact that half the bursts of Cerenkov radiation were pointing out of the front of the eye, and only the half which point towards the retina could be detected. So, on this calculation, if the entire population of the Earth had its eyes closed, and was dark-adapted at the moment the wave of neutrinos passed through the vitreous material in the 100 000 tonnes of human eyeballs worldwide, perhaps 100 people might have noticed a sparkle of Cerenkov radiation. They would have to have picked it out from the Cerenkov radiation induced by the cosmic ray background of naturally occurring charged particles which penetrate our eyes all the time.

A more realistic calculation takes into account the detailed mechanism of the way the human eye detects photons. Fazio, Jelley and Charman (1970) calculated whether flashes of light which were seen from translunar orbits by Apollo astronauts Buzz Aldrin, Neil Armstrong, Charles Conrad, James Lovell, Jack Swigert and Fred Haise could have been Cerenkov radiation from cosmic rays. Light excites the eye when 4–10 photons or more are caught within the basic sensitive unit of the retina – a so-called ganglion cell. There is a threshold of about four photons per detector element needed to trigger a response. The angle of the cone of light from Cerenkov radiation means that the light emitted when a neutrino interacts with a proton at the front of the eye is spread over a large portion of the retina. The photons are spread too thinly over the numerous ganglion cells to trigger them. Only the third millimetre of track

Fig. 30. On the left a cross-section through the retina shows the essential cells; on the right is the track of a relativistic electron (or positron) accelerated in the vitreous humour of the eyeball by a supernova neutrino. The Cerenkov radiation from the electron radiates over a cone of angle 42°, and activates numerous receptor cells unless produced close to the retina.

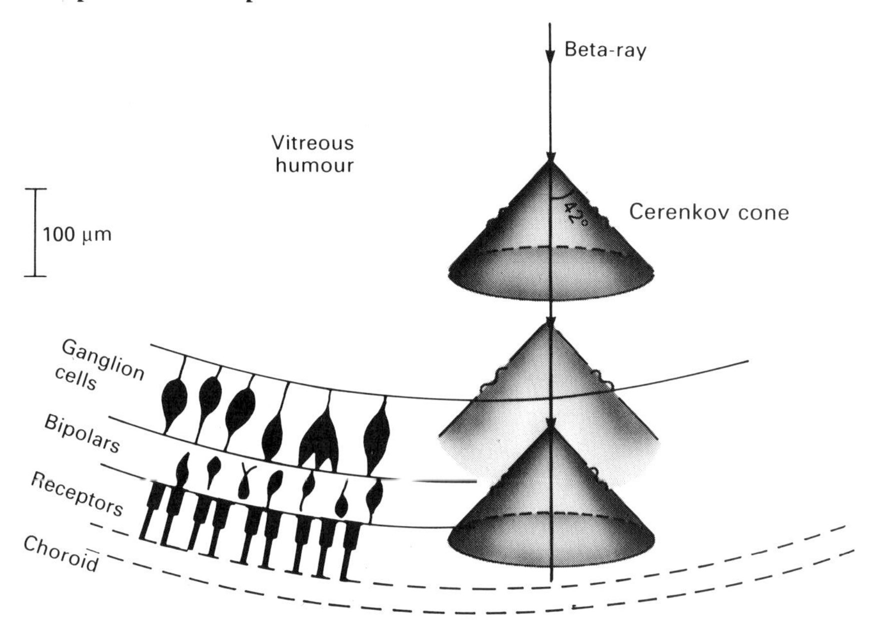

nearest the retina emits radiation which hits one cell, and this is generally of an intensity of only three photons, not enough infallibly to excite the eye. There is a detectable flash in only one impact in a 100–1000 or so. In 100 impacts, at most one person saw the Cerenkov radiation from the supernova's neutrinos. If this unique person exists, he gives a new meaning to the word 'star-struck'.

Reactions to the neutrino events

IMB and Kamiokande saw neutrino bursts at the same time, within the uncertainty of the inaccurately set Kamioka clock. The two detectors are similar and saw neutrinos of similar numbers and energies.

It is not credible that detectors in different continents should suffer some malfunction or 'glitch' at the same moment which could mimic a cosmic event so consistently, and everyone agrees that the IMB/Kamiokande neutrino pulse was real. Everyone also agrees that it is far and away most likely that the pulse came from the supernova in the LMC. It strains everybody's credulity to imagine that it is chance that the only cosmic neutrino pulse ever seen should happen the day before the discovery of the brightest supernova for nearly 400 years.

Fig. 31. The energy of the neutrinos detected by the Kamiokande (open squares) and IMB detectors (filled squares) is plotted as a single neutrino pulse in this combined plot. Most neutrinos arrived within the first second or two but there was a lower energy tail to the pulse which lasted more than 10 s.

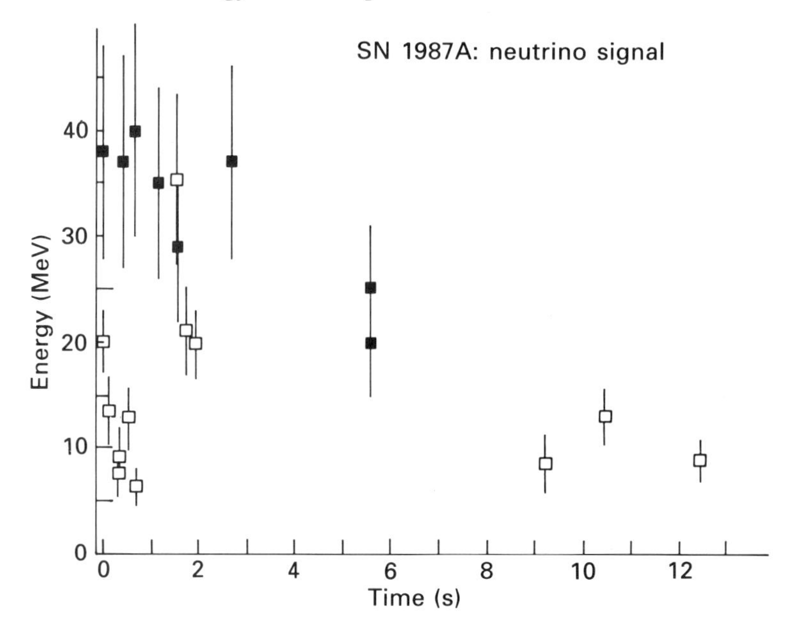

If the IMB/Kamiokande pulse is real, did Mt Blanc and Baksan see it too – and if not, why not? These observatories are smaller than IMB and Kamiokande but are more sensitive to the more abundant low energy neutrinos. Taking this into account it might be expected that Mt Blanc should detect an average of 1.5 events, and most likely no, one, two or three events, at the same time as the IMB pulse. Mt Blanc detected events at 07:36:00.5 and 07:36:18.9 UT (Aglietta *et al.* 1987b); these follow the IMB pulse by 20–40 s and they may be background events – but they might be from SN 1987A. There is no problem here; 'up to two events', which is what was observed, is consistent with 'between none and three', which is what would be expected.

Fig. 32. Comparison of the four neutrino detectors and the events which they detected. The numbers of events detected differ by only a factor of 2, but the detector masses differ by a factor of 50; on the other hand the smaller detectors can see the (presumably) more numerous less energetic neutrinos. It is possible (just) to fit these data together into a consistent pattern. After a diagram by A. Wolfendale.

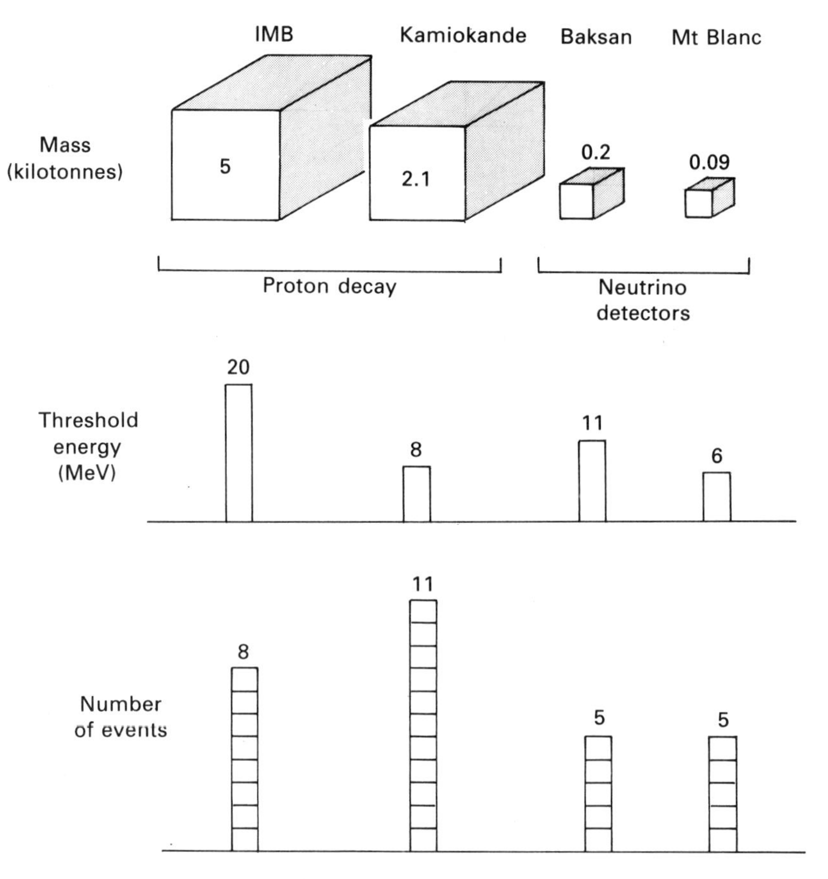

The Baksan detections are more problematical: it too should have seen no, one or two neutrinos at 07:35:40 UT. It saw five – three times more neutrinos than expected from the IMB/Kamioka pulse (Alexayev *et al.* 1987) – and they also arrived 30–40 s after the accurately timed IMB neutrinos. It seems that some or all were background events.

It thus seems possible to develop a consistent account about the neutrino pulse at 07:35 UT, and what all four observatories saw at the time – the two big observatories clearly saw the same pulse and the two small ones might or might not have been big enough to see it too, and maybe they did or maybe they did not.

What about the pulse at 02:52 UT? This was seen in the Mt Blanc observatory, perhaps or perhaps not (one event) in Baksan, but nothing in IMB. In Kamiokande there was a single event which lay below the energy cut-off which defines the boundary between definitely real higher energy and lower energy events which are probably background. Referring to the claim of an observation by the Mt Blanc observatory, Hirata *et al.* (1987) say that 'no evidence for an event burst outside of statistics is present in our data at 02:52 . . .'.

Fig. 33. The signal seen in the Kamioka detector at the time of the supernova signal claimed by the Mt Blanc neutrino observatory. The event at 12 MeV at a time between 4 and 5 min in this diagram occurred on 1987 February 02:52:40 UT, during the Mt Blanc pulse. The event at a time of 11 min was 6 min afterwards. The Japanese experimenters concluded that 'no evidence for an event burst outside of statistics is present' in this data.

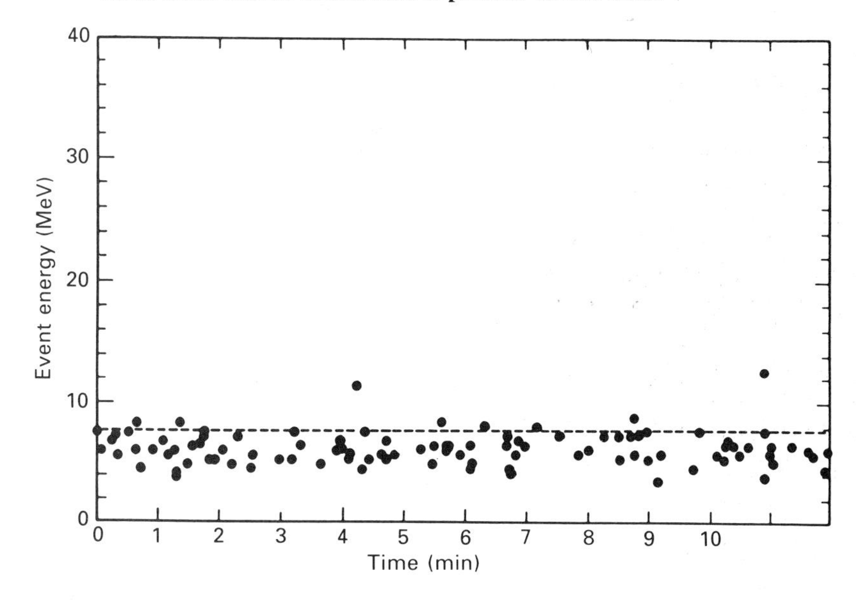

Now, it is more difficult to explain why something was seen in the small instrument but not in the larger. But there is a possible explanation which forces us to make assumptions about the spectrum of the neutrinos. If most neutrinos in the 02:52 UT burst were low energy, then the Mt Blanc Observatory (which sees neutrinos down to 6 MeV) could see several while Baksan, Kamiokande and IMB (10, 8 and 20 MeV respectively) saw fewer. So there are possible differences in detector efficiency and response as a function of neutrino energy which make it possible that the Mt Blanc observation could remain undetected by the other observatories.

If this is right then there were not one but *two* neutrino bursts on the day before the discovery of the supernova in the LMC.

Some astronomers and physicists have found this much more difficult to accept than others. Some regard the Mt Blanc observation as 'unlucky': a chance event – an experimental glitch perhaps, or a statistical fluke – fortuitously on the same day as the explosion of SN 1987A. Discussions have been, to quote an intended euphemism, 'vigorous' (Helfand 1987). It is a debate which has gone beyond the dispassionate evaluation of the technical, logical and scientific facts, in the manner believed by non-scientists to be typical of scientific discussion. The debate has proved that scientists are human beings who use personal judgement, who get angry, who compete for recognition, who have to interpret data which are not totally clear, who have to evaluate credibility and who have biases and prejudices. At its crudest the argument was couched in terms of nationalism.

As examples of the genuine differences of opinion about the interpretation of evidence here are two examples. Helfand (1987) quotes Thomas Walsh (Minnesota) as saying in a report of discussion at a workshop in early June in 1987: 'The most important problem [with the Mt Blanc results] is the lack of a consistent story on their statistics; background bursts of the magnitude seen on February 23 have been variously reported as occurring once a month, once every five months and once every two years.' Carlo Castagnoli for Mt Blanc countered that 'since the very beginning we have reported the same value for the imitation rate of our burst by background: 0.7 per year.'

In another difference of opinion, in 1987 July, at the ESO workshop, P. Galeotti described the Mt Blanc burst and showed, with a graph of the data, that (as explained above) Kamiokande detected a single event at the same time, saying that this was consistent with the known detector efficiencies. Masa-Toshi Koshiba of the Kamioka collaboration was infuriated by Galeotti's display of the Kamiokande data in support of the Mt Blanc observation, which the Kamioka collaboration was on record as not confirming. The event was

Top. Star trails curve in the southern sky above the dome of the AAT on Siding Spring Mountain, Coonabarabran.

Bottom. The Kuiper Airborne Observatory is in an adapted C-141 military transport, here flying over San Francisco.

Top. Ian Shelton, codiscoverer of the supernova, took this 3-hour photograph of it in excellent atmospheric conditions with the University of Toronto's 24 in telescope at Las Campanas on 1987 March 26. © University of Toronto.

Bottom. A 30 minute exposure on Fujichrome by Brian Carter of the SAAO shows the supernova orange on 1987 April 30, south west of the red 30 Doradus Nebula, here exhibiting its 'tarantula' shape.

Top. David Malin's 'before' and 'after' pictures of the supernova, made by combining pictures with the AAT, emphasise the red colour of the nebulae in the region and of the supernova (orange halo to the saturated pseudo-white bright image).

Bottom. The 30 Doradus Nebula is the brightest nebula in the LMC and is a region of bright stars in clusters and of dust and nebulae. The supernova occurred in this region and its progenitor was a massive star like all the brighter stars visible in this picture. Photo by David Malin © AAT Board 1987.

Top. Sunrise at the Las Campanas mountain, Chile. In the centre of the ridge is the University of Toronto observatory, with the 10-in Bruce telescope in the barn-like building.

Bottom. The European Southern Observatory, La Silla, Chile. One of the world's greatest arrays of telescopes is dominated by the 3.6 m telescope building at the rear. © ESO.

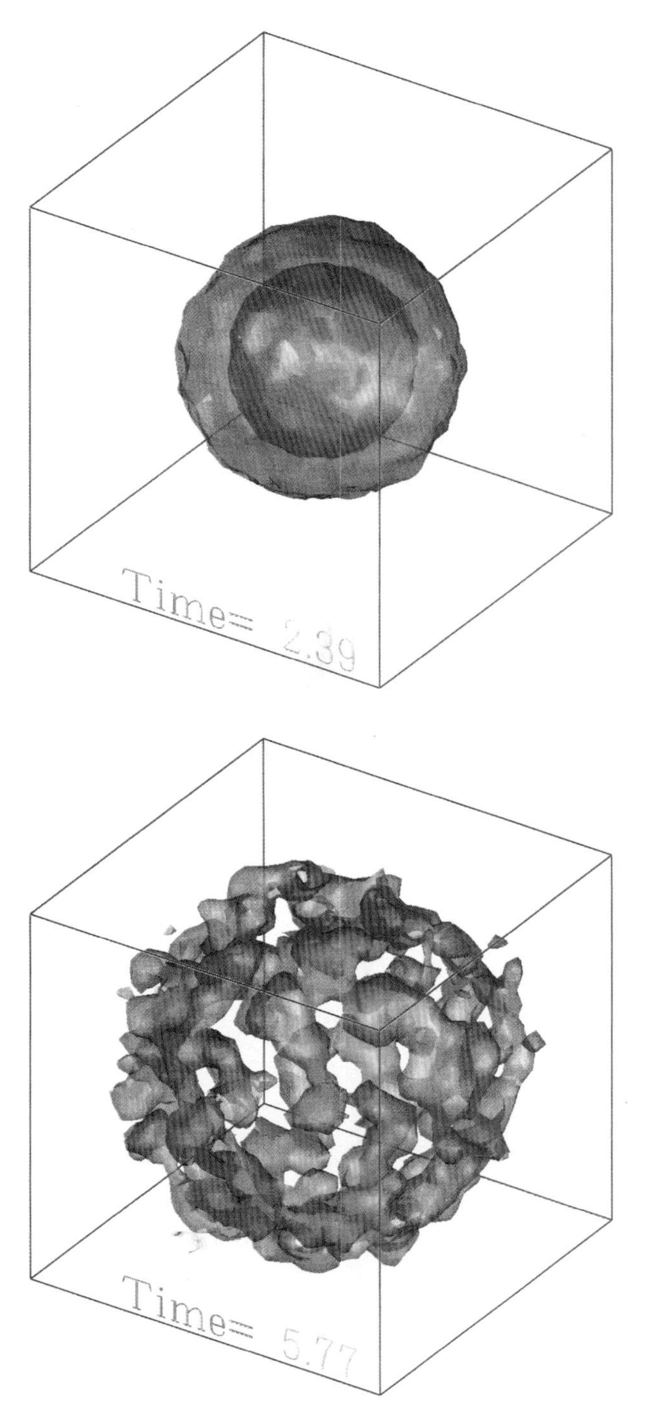

Computer simulations by M. Nagasawa, T. Nakamura and S. M. Miyama of Kyoto University of the expansion of the core of the supernova show how in a few seconds the core fragments into lumps which stream radially outwards. These lumps penetrated into the envelope of the supernova and distributed radioactive nickel towards the surface of the star.

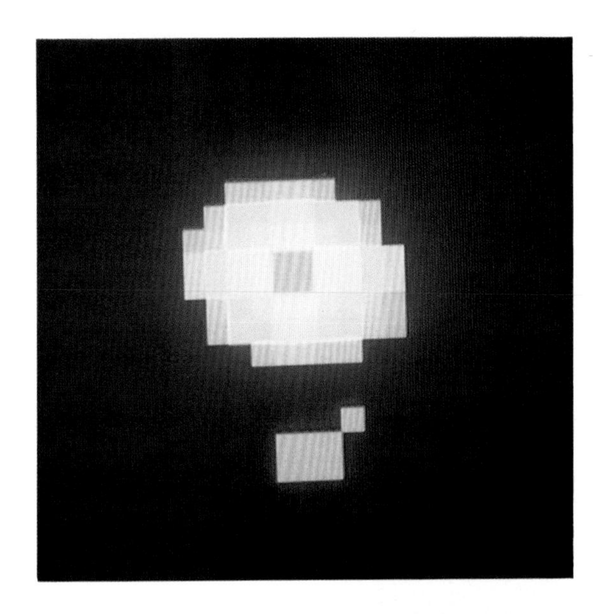

Images from the Center for Astrophysics' speckle interferometer used on the 4m CTIO telescope (courtesy of M. Karovska, CfA). North is at the top, east to the left.

Top left. In 1987 March, the supernova is accompanied by the Mystery Spot.

Top right. In 1988 March, the supernova is elliptical, but the Mystery Spot has disappeared.

Bottom. Another star observed for comparison is completely symmetrical.

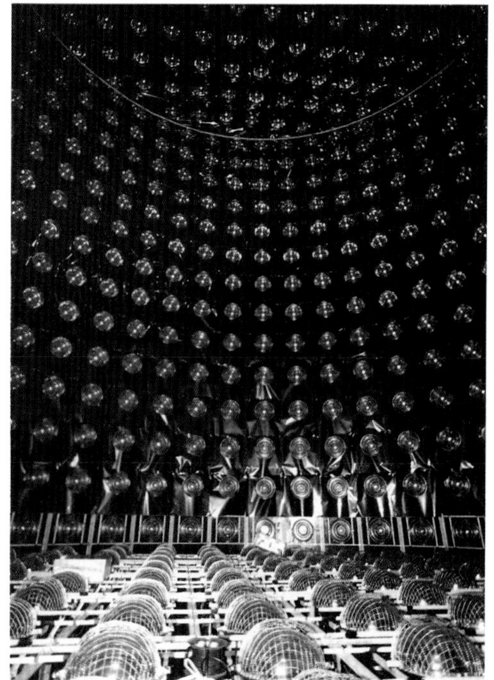

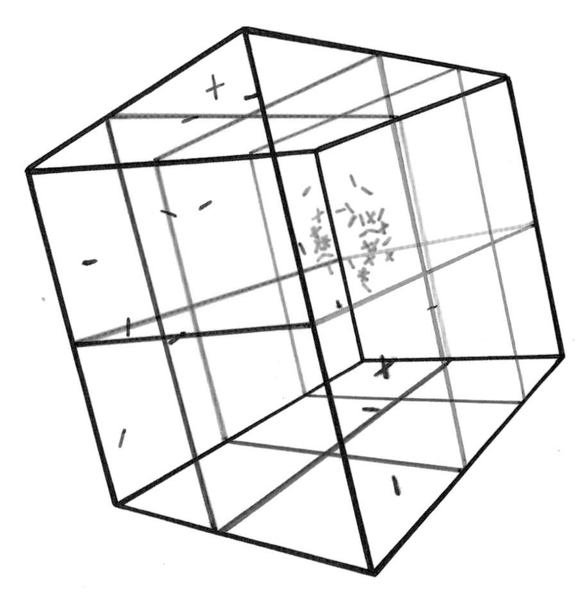

Top left. An image of SN 1987A as taken with the Fine Error Sensor (FES) of the IUE satellite. The cross indicates the supernova, which far outshines the other stars in the same region. Analysing this image during an observation of the supernova, the FES holds the IUE spacecraft steady against the drift which is produced by the solar wind.

Top right. The inside of a neutrino observatory is a tank covered with hemispherical photomultiplier tubes, like the facets of an insect's eye. They detect the Cerenkov radiation emitted after neutrinos interact with electrons or protons in the water with which the tank is filled.

Bottom. Diagram from a computer display of a Cerenkov event from one of the neutrinos from SN 1987A detected by the IMB detector. The box structure represents the tank of water and the marks on the walls represent photomultiplier tubes which have been fired by the Cerenkov radiation. The elliptical patch of activated photocubes cuts across the cone of Cerenkov radiation.

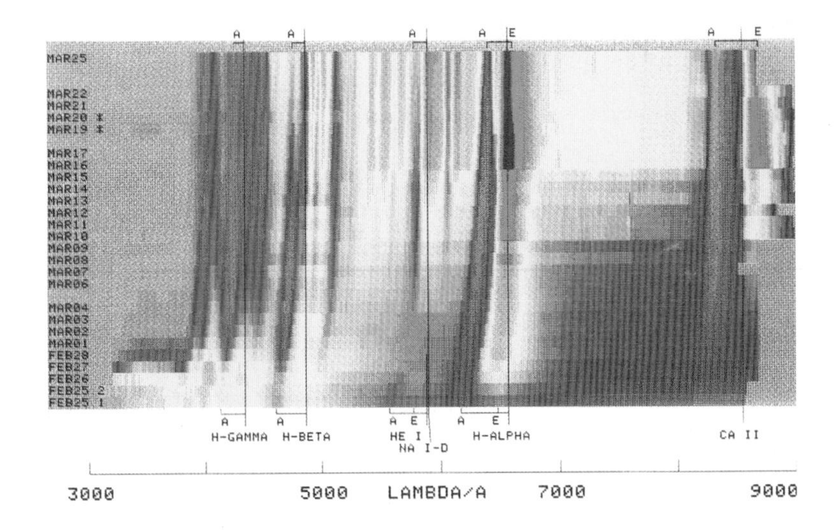

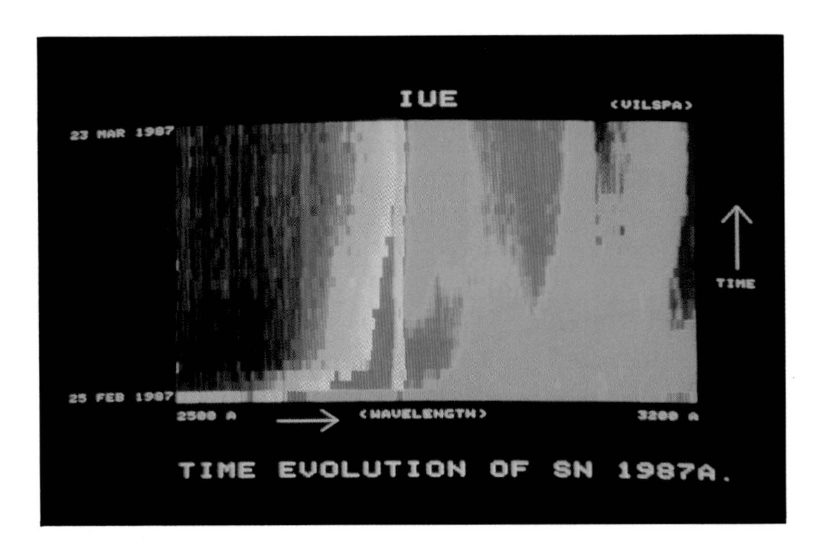

Top. Colour-coded plot of optical spectra of SN 1987A as obtained by R W Hanuschik and J Dachs with the University of Bochum telescope at ESO shows how the spectrum changed over the first month (time runs day by day from bottom to top, wavelength from left to right). Very noticeable is the marked fading of the ultraviolet part of the spectrum in the first five days of the outburst (bottom left). In early March more and more absorption features (vertical stripes) began to appear as the supernova cooled. The drift and curvature of the emission at H-alpha (prominent green and blue near-vertical strip to the right of centre) shows how the spectrum progressively came from the slower-moving deeper parts of the supernova as it expanded.

Bottom. Similar colour-coded plot of the ultraviolet spectrum of SN 1987A as obtained by IUE. Courtesy of W. Wamsteker.

background and should not be used in support of the Mt Blanc result, he said. 'Using data is the responsibility of the experimenter,' he claimed. Introducing his own talk later, he remarked, apparently ironically, 'I should be happy that our data was quoted already so many times.'

The underlying difficulty which raises the temperature of the debate is, not simply that two neutrino bursts were apparently seen and this is difficult to understand in itself, but that, whereas the Mt Blanc pulse at 02:52 UT throws up difficulties in interpretation by physicists, the 07:35 pulse of neutrinos so nicely fits with the standard theory of supernovae (Krauss 1987, Burrows and Lattimer 1987) that physicists are almost embarrassed. 'I am surprised that the simple theory holds up so well,' wrote Burrows (1987b) in relation to the IMB/Kamiokande pulse of neutrinos, 'and suspect that in this I am not alone.'

What does the intercomparison of the observations by the four different neutrino observatories consist of? It is actually a rather difficult exercise and fully deserves the name of 'neutrino spectroscopy' (Krauss 1987). It is too complicated to go into here but it consists of the following stages.

First, the numbers and energies of the neutrinos emitted from the supernova must be calculated. Each process must be taken into account and due attention paid to the various flavours of neutrinos created.

Second, the way in which each detector views the various neutrinos must be fed in: for each flavour the various ways in which the neutrino gives rise to an

Fig. 34. The modes of production of neutrinos and their detection have to be fitted together with the detector characteristics to make a self-consistent pattern of 'neutrino spectroscopy' for the score of neutrinos actually seen.

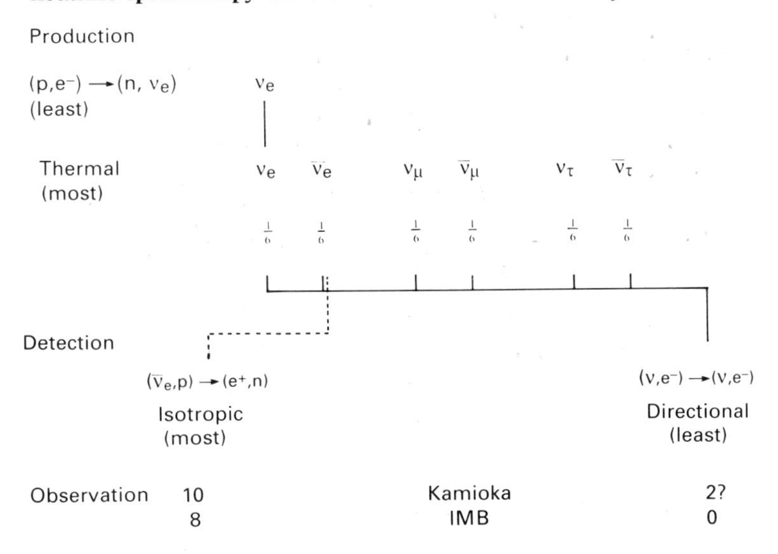

event must be calculated. The calculation must take account of the relative numbers of protons and electrons in each detector. The calculations depend on the energy of the neutrinos, some detectors recording more energetic neutrinos than others.

Then this must be compared with the observations paying attention to the way that the observatories recorded, or did not record, the direction of the neutrinos and what that implies about the various sorts.

Finally, all this has to be done with regard to the statistical uncertainties. If eight neutrinos have been recorded then the laws of statistics tell us that this number is subject to an error of 35%, and there is a wide band of uncertainty. If we divide these eight into some which are high energy and some which are low, or some which were from the LMC and some which were not, then the number of neutrinos in the subdivisions is even smaller than eight and therefore even more uncertain.

Clearly there is room for interpretation in the results.

The energy from each event

The real problem of the incompatibility between the two neutrino bursts which were detected is the energy which each implies was released at the supernova. The Kamiokande/IMB pulse implies that the total neutrino luminosity released in SN 1987A in the LMC was about 3×10^{46} J. As we calculated before, this is equal to the total annihilation of 0.1 the mass of the Sun, which, considering that there was an object of a couple of solar masses involved in the core collapse is amazingly efficient in converting mass to released energy (5%). (Compare an H-bomb in this respect, or the Sun, at 0.7%.) At the same time, this is in accordance with the theoretical expectation for the formation of a neutron star and the release of its gravitational binding energy.

The Mt Blanc pulse implies that several times this energy was released: $(12-60) \times 10^{46}$ J in electron anti-neutrinos alone (Aglietta *et al.* 1987), and about six times this energy in all neutrino species. At the lower limit, this is equivalent to the annihilation of 2.5 times the mass of the Sun, i.e. the total annihilation of the core of Sk − 69 202 (100% efficient). To make this credible there must be something wrong with the theoretical calculations, e.g. the inference that there are six neutrino species which all are created in equal numbers in the collapse of the core (Galeotti 1987). This is not ruled out, but few scientists believe that it is likely. Indeed some infer by comparing the Kamiokande/IMB pulse with the theory of the formation of neutron stars that there are no more than the six species of neutrinos already known, certainly less than eight.

Table 18. *Rest masses of all flavours of neutrinos.*

mass of electron neutrino	<18 eV (Fritschi *et al.* 1986)
mass of muon neutrino	<0.25 MeV (Boris *et al.* 1985)
mass of tauon neutrino	<70 MeV (Particle Data Group 1986)

The mass of neutrinos

These neutrinos from the supernova in the LMC are the first seen from outside the solar system and the first directly attributable to an astronomical event. Thus SN 1987A is a milestone in the subject of neutrino astronomy. It is not surprising that their detection sparked great interest amongst physicists. According to an editor of *Physical Review Letters*, in the five months after the discovery of the supernova the journal had received 36 papers interpreting various physical aspects of the supernova (Brown 1987). Many of them were on the possibility of using the neutrino burst from the supernova at least to set new and stringent limits to the masses of neutrinos, and perhaps to determine them.

This would be of great interest in debates about the density of the Universe. About 85% of the mass of the Universe is unaccounted for and exists in the form of 'dark matter' (Disney 1984; Kormendy and Knapp 1987). Some of this mass may reside in the form of the neutrinos created in the Big Bang at the start of the Universe. The present density of neutrinos from the Big Bang is 110 neutrinos per cubic centimetre. If the mass of all flavours of neutrinos is the same and greater than 15 eV, then the Universe is closed, i.e. the expansion of the Universe will eventually cease. The Universe will collapse back in on itself in what is colloquially known as the Big Crunch.

We have already looked at the laboratory upper limits to the electron neutrino rest masses in Chapter 7. The upper limits to the rest masses of the other flavours are included in Table 18. The upper limit to the mass of the electron neutrino is tantalising for cosmologists, while the upper limits to the two other flavours are useless to them; thus cosmologists have become very interested in the further results from the LMC supernova.

A pulse of neutrinos (or a pulse of any other particles) is lengthened into a train as they travel through space because individual particles travel at different speeds v (less, by varying amounts, than the speed of light, c, if the particles have a non-zero rest mass). The relativistic energy, E, of a neutrino of rest mass, m_0, is

$$E = mc^2$$
$$= m_0c^2/\sqrt{[1+(v/c)^2]},$$

so

$$v = c\sqrt{[1-(m_0/m)^2]}$$
$$= c\sqrt{[1-(m_0c^2/E)^2]}.$$

The time of arrival t of the particles after they travel a distance d is

$$t = d/v$$
$$= d/c + m_0{}^2c^3/2E^2 + \ldots.$$

In principle therefore the time of the detection of the neutrinos as a function of $1/E^2$ should show a straight line whose slope is proportional to $m_0{}^2$. Several groups of physicists wrote analyses of the supernova's neutrinos' rest masses,

Fig. 35. The energy, E, of the two dozen neutrinos observed at Kamiokande and IMB (different symbols) can be plotted as $1/E^2$ versus arrival time. If the neutrinos were emitted simultaneously and their arrival times were dispersed by the contribution of their rest mass to their energy, then the plot should be a straight line whose slope was proportional to $m_0{}^2$. Three straight lines show the range of fits which might accommodate three flavours of neutrinos (whose masses are expressed as energy equivalents of 3.4, 17 and 28 eV respectively). However, the diagram is inconsistent because the later-arriving neutrinos on this interpretation must be muon and tauon neutrinos, which must produce electrons whose trajectories point away from the LMC: they did not. Hence this is not the explanation for the spread of the neutrino arrival times in a tail away from the bunch which arrived in the first 2 s.

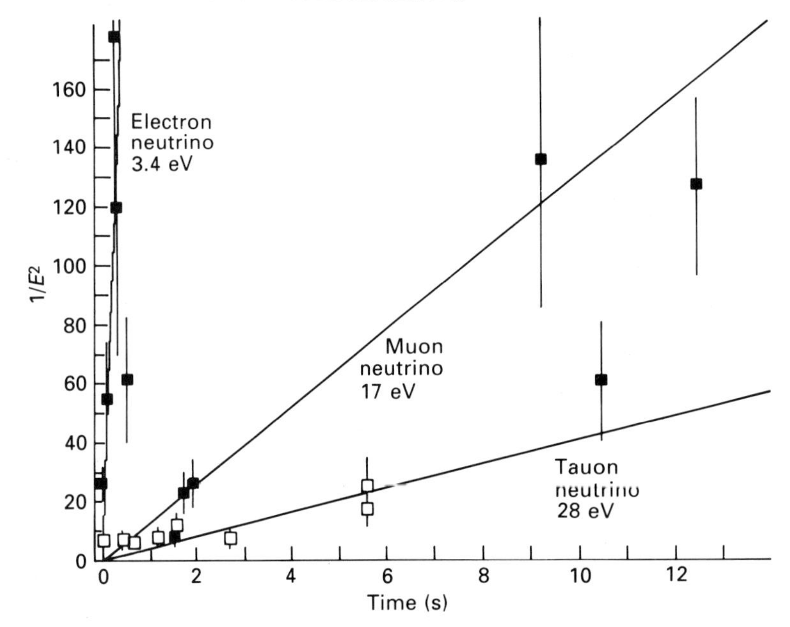

based on these ideas (ten papers were submitted in the first two months after the supernova, to *Physical Review Letters* alone); alas, in the rush to publish several points were overlooked.

Since there are three flavours of neutrinos it might be possible, in principle, to find three straight lines in the data, if the neutrinos were all represented in it. But, as indicated above, most of the neutrinos detected from the supernova were electron anti-neutrinos. In Kamiokande, which has the most complete record of the directions of the neutrinos and therefore their flavours only two are thought to be candidates to be muon or tauon neutrinos. These are, in fact, the first two detected and therefore travelled here the quickest and should be thought of as the least massive, on the above analysis. Graphs such as Figure 35, which show the muon and tauon neutrinos as the more massive (as the upper limits of Table 18 allow) are internally inconsistent with the actual data.

A crucial assumption is that the neutrinos were all emitted from the supernova at the same time, or at least in a much shorter time than the spread of arrival times (typically 10 s) observed in the neutrino observatories.

The theory of the diffusion of neutrinos out of the supernova is, as can be imagined, not well defined. But it appears to suggest that the short burst of neutrinos created in the collapse to a neutron star is diffused to a lengthier pulse in its transmission through the star material (Burrows and Lattimer 1987). Only after the diffusion and the emission of the neutrinos from the 'neutrino-sphere' do the neutrinos set out unhindered across space.

The calculations are based on the known and calculated properties of neutrinos (electroweak theory, including neutral currents) and the so-called standard model of the collapse of a supernova core. The calculations suggest that the diffusion takes place over a second or two, with a tail of the late-leaving neutrinos which lengthen the pulse for up to 10 s – the time actually observed. There is not much room for extra lengthening of the pulse by the relativistic dispersion due to a finite rest mass of the neutrinos.

At the moment therefore all that can be said with certainty is that the data suggest an upper limit to the mass of all the neutrinos detected of about 10 eV. Neutrinos are probably not massive enough to close the Universe.

The lifetime of neutrinos

Another fundamental fact about neutrinos which the supernova reveals is their mere detection after they have travelled for 170 000 years.

In experiments, physicists have detected neutrinos from nuclear reactors, particle accelerators and cosmic rays. All of these terrestrial sources lie within a maximum flight distance set by the diameter of the Earth. Neutrinos move nearly at the speed of light, and so the ones detected in laboratory experiments from terrestrial sources were created no longer than 0.03 s before detection. We know, therefore, that they survive this long – but could they decay on a longer timescale?

One reason why this is an important question is in relation to the solar neutrino problem. If neutrinos decay over the 8 min light travel time from the Sun then this may be considered an explanation for the deficiency in observed solar neutrinos.

The observation of neutrinos from the LMC after they have travelled 170 000 years rules out neutrino decay over the 8 min light travel time to explain the deficiency of neutrinos from the Sun. Suppose they decay to a noticeable degree in 8 min – say 1% disappear in that time. Then their lifetime τ is such that

$$\exp(-8\ \text{mins}/\tau) = 0.99$$

This would give

$$\tau = -8/\log_e(0.99)\ \text{min} = 795\ \text{min}.$$

Fig. 36. Kamioka and IMB data are histogrammed versus time to show the sharp pulse of neutrinos which arrived in the first 2 s, followed by a tail of later events which followed up to 12 s later. Superposed are the results of two model calculations, which fit the data reasonably well. The model calculations include the effects of neutrino diffusion. Diagram after A. Burrows.

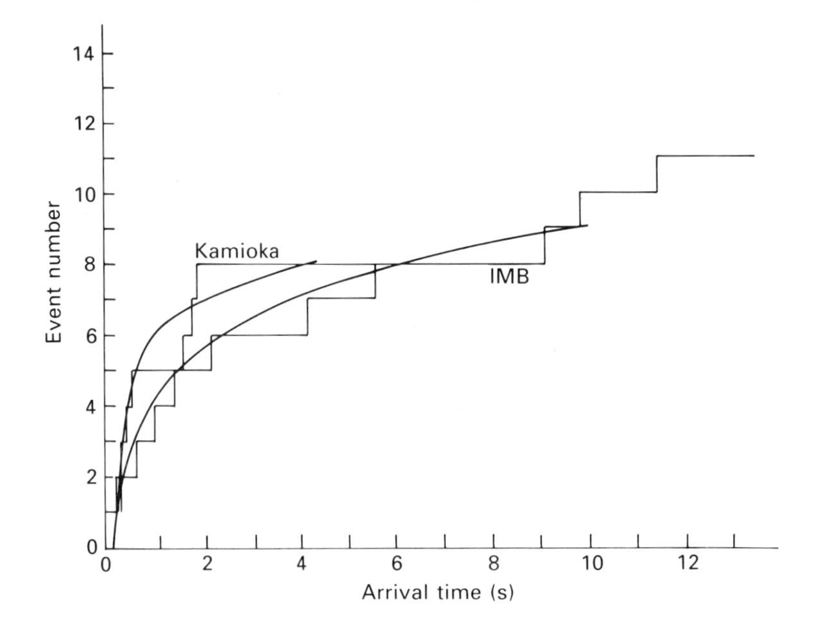

So the fraction of neutrinos which survive after 170 000 years $= 9 \times 10^{10}$ min would be

$$f = \exp(-9 \times 10^{10}/795).$$

This number is indistinguishable from zero!

The observation of the LMC neutrinos sets a lower limit to the lifetime of neutrinos of $1.7 \times 10^5/\gamma$ years. The factor, γ, is the time dilation factor and is related to the neutrinos' speed. There are clocks in the neutrinos, of some kind or another, which make the neutrino decay on a certain timescale. The clocks are moving quickly with the neutrinos and run slow relative to ours in accordance with the special theory of relativity. The faster the neutrinos move, the slower their clocks run and the longer the neutrinos appear to last to us, regardless of what the neutrinos themselves believe their lifetime to be according to their perception of their own clocks. In fact, if neutrinos are exactly of zero mass and travel exactly at the speed of light, then they last for ever, so far as we are concerned, just like light itself. This can be seen from the expression for γ as the ratio of the neutrino energy to its rest-mass energy:

$$\gamma = E/m_0c^2.$$

If the rest mass of the neutrino is zero then γ is infinite; regardless of the distance of the LMC – 170 000 light-years or megaparsecs – the neutrinos will get here without decaying, just as photons do. This again would mean that the solar neutrino problem is not just going to fade away!

SN 1987A as an individual supernova

All supernovae are initiated by the gravitational collapse of a star. The part of the star which collapses is the core of a massive star in the case of a Type II supernova, or, it is believed, a white dwarf in the case of a Type Ia. There may be other types of stars which collapse, as well. The end result, or, at least, an intermediate stage, of the collapse is a neutron star. All neutron stars seem to be very much the same mass, 1.4 solar masses or so, and the same size, say radius 50 km. The radius is much smaller than the progenitor star. Thus, more or less independently of what kind of star collapsed, the gravitational potential energy released in the collapse is always the same. This means that in gross terms all supernovae start off the same way, and that for example the neutrino bursts from all Type II supernovae would be very similar. This makes it possible to conceive of a 'standard model' by which to examine the neutrino burst of 1987 February 23.

But the environment surrounding the release of this energy in the core varies from supernova to supernova: the progenitor stars are different sizes, different masses, different compositions, different mass distributions; they are single, or in multiple stars; they are stripped of any envelope at all or, by contrast, have extended atmospheres and circumstellar envelopes, with and without strong magnetic fields.

The different packages make the differences from supernova to supernova.

This is not an unusual concept in principle. In the same way that supernovae are at once the same in general and different in particular, all human beings are biologically the same but different one from another.

Human beings are members of a single species and a typical human can be described in general quite accurately: 50–100 kg in mass, 1.5–2 metres high, water-soluble carbon-based biochemistry, a bipedal mammalian with binocular vision connected to sophisticated image processing system, self-replication by helical DNA duplication within 15–40 years, death by 5×10^9 s etc.

The interesting differences, which make all the difference (*vive la difference!*), are matters of packaging: colour, shape, nationality, personality, sense of humour, talents and abilities, etc. From such 'superficial' differences come all the important and fascinating endeavours of human history, such as art, science, war, sociology, love, religion and belief, commerce, industry . . .. To understand the human race, it is necessary to know individuals in it. It is just so with supernovae.

In the remaining chapters, we look at SN 1987A as a particular individual.

9 The supernova expands

The core collapse inside Sk – 69 202 released large numbers of neutrinos; some of these neutrinos interacted with the material of the body of Sk – 69 202. In addition, the core collapse generated a shock wave which travelled out and up through the body of the star, heating it throughout. Both these processes deposited energy and momentum in the star's envelope.

The fraction of gravitational potential energy deposited in the envelope was very small (much less than 1% – see Table 19), but was enough to produce the visible effects of the supernova. Most of the energy went into the neutrino flux. It is chastening to realise that, however important astronomers thought supernovae were, basing their belief on the light from them and the kinetic energy of the expansion of the material, they were simply ignorant of the true magnitude of a supernova explosion until they had learnt of the neutrino emission. If astronomers weighted their studies according to the importance of the energy emitted in the phenomenon, then this book would be nearly all about neutrinos from the supernova and only a few pages would be devoted to everything else!

Nonetheless the motion and radiation of supernovae are manifestations of their expansion – and the energy in these forms is what called attention to supernovae in the first place!

Expansion of SN 1987A

As the shock from the core collapse broke out from the surface of Sk – 69 202, its envelope began to expand outwards. At first this expansion was homologous: the scale of the distribution of the matter throughout the star

Table 19. *Expansion energy of the supernova*

neutrino energy (IMB/Kamiokande pulse)	3×10^{53} erg
initial kinetic energy of expansion	10^{51} erg
initial radiative energy	3×10^{46} erg
total radiative energy	7×10^{48} erg

changed as the star expanded, but the relative distribution of material was not altered – just as if a photographic zoom lens was progressively enlarging the star. The outer parts were expanding fastest, the inner parts slowest, with a gradation between. Since astronomers have ways of estimating speeds of material, they can interpret the speeds in terms of the depth of various layers in the supernova.

The outflow of the atmosphere of the star, driven outwards and upwards by the surface below, was seen in the spectra of the supernova.

If the thin atmosphere was viewed in isolation it would show in the optical

Fig. 37. Energy flow paths in the core collapse of supernova 1987A, summarising Chapters 6 and 8 of this book, Chapter 9 starts in the lower right hand corner.

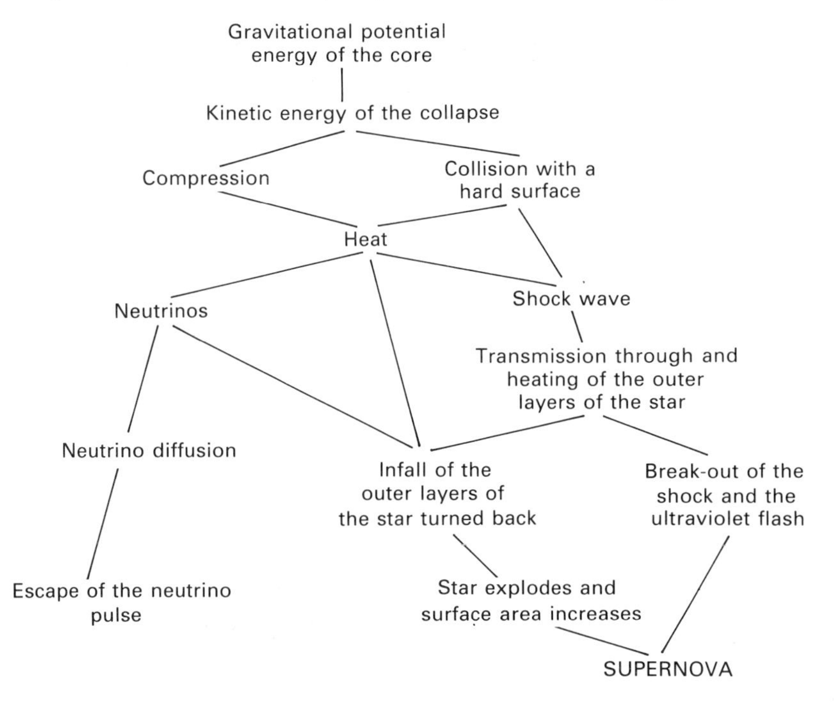

spectrum as emission lines which would be broad because the atmosphere flowed outwards radially from the centre of the star, projected at various angles onto our line of sight. In fact the atmosphere surrounded the supernova itself, which was optically thick and formed a hot surface below the thin atmosphere.

Hence, the rear part of the atmosphere, flowing away, was hidden behind the surface layers of the supernova. Thus the red tail of an emission line profile was blocked from reaching us. We could, however, see the emission from the sideways, outflowing part of the atmosphere – this emission was centred on the rest wavelength of the spectral line.

Moreover, the front part of the atmosphere, approaching us at maximum velocity, was seen in projection against the hotter surface layers. Rather than appearing as an emission line, the front part of the atmosphere manifested itself as an absorption line. Because it was from the approaching part of the

Fig. 38. The Kamioka and IMB observations of neutrinos require a very much smaller amount of energy than the Mt Blanc observation. Even so, the neutrino energy in the production of SN 1987A is far larger than all the energy which we otherwise 'see' from the supernova. Diagram after one by A. Wolfendale.

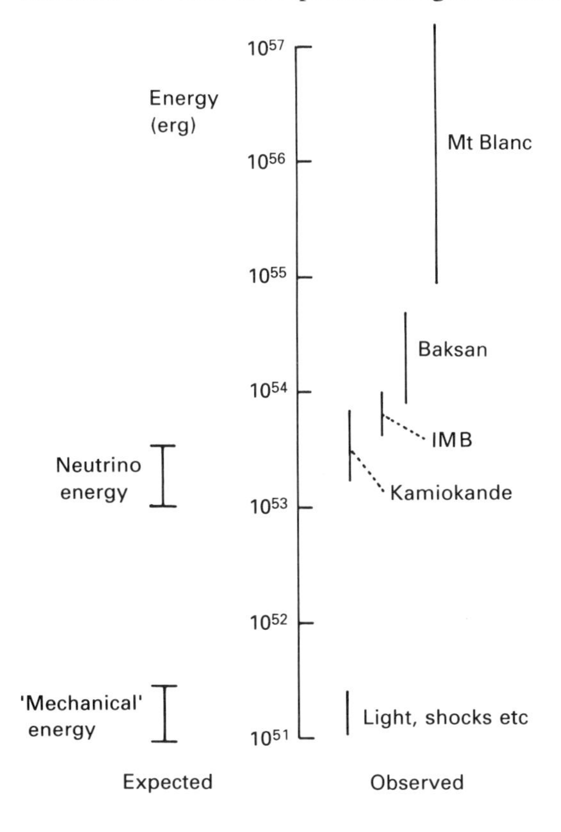

Fig. 39. The 'P Cygni' profile is a natural consequence of an expanding atmosphere producing line radiation above a 'photosphere' of continuum radiation produced where the star is optically thick. The thick part of the star hides the rear (receding, red shifted) part of the atmosphere, while the front (approaching, blue shifted) part of the atmosphere appears as an absorption line. The sideways moving parts of the atmosphere produce the emission line. The offset of the absorption dip from the rest wavelength, λ_0, of the spectral line measures the expansion speed, V, of the atmosphere above the photosphere.

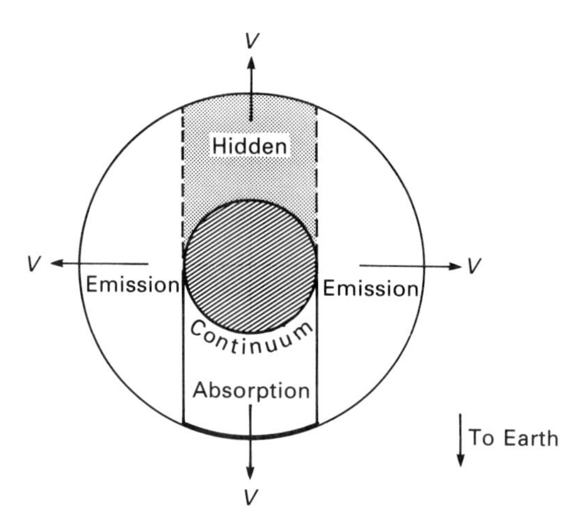

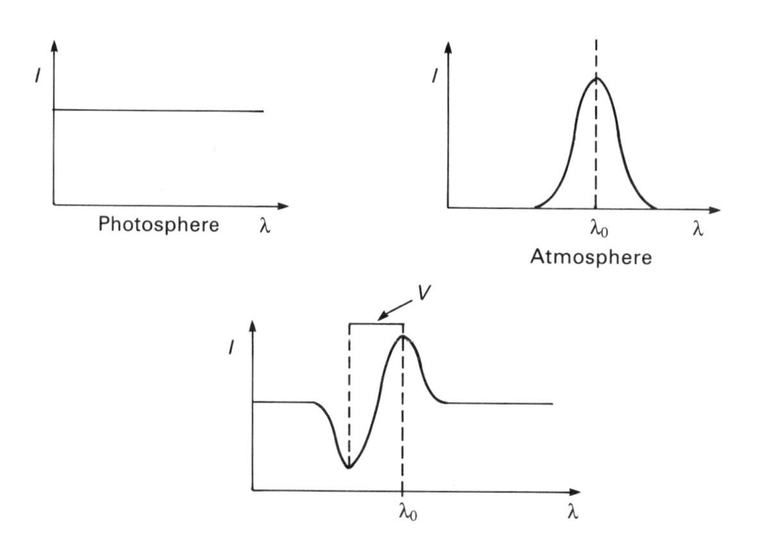

expanding atmosphere, the absorption line was blue shifted from the rest wavelength of the spectral line.

These geometrical effects gave a characteristic S-shaped signature to the hydrogen lines, which astronomers call a P Cygni profile. The star P Cygni has such an S-shaped signature to its emission line profiles because its atmosphere is outflowing, although not so fast as SN 1987A!

The speed of the outflow of the thin part of the atmosphere of the supernova is given by the displacement of the absorption dip from the rest wavelength of the spectral line. The displacement, converted to a velocity by the Doppler formula, is the outflow speed of the layer where the absorption line forms. There is no consensus amongst astronomers whether to quote the displacement of the centre of the dip, which represents an average of the outflow speed of the thin part of the atmosphere, or the extreme displacement of the furthest traceable blue part of the dip, which represents the outflow speed of the topmost part of the atmosphere. Indeed, it is not really meaningful to talk about 'the' expansion speed of the atmosphere since different spectral lines have different P Cygni profiles. This is because the spectral lines are formed above different surfaces in the atmosphere (Bildsten and Wang 1988) and these different surfaces are expanding at different speeds, the topmost expanding faster than the lower depths.

The supernova was not blowing up like a balloon, but like a puff of smoke.

At the SAAO, spectra of the supernova have been obtained practically daily since the middle of the night of 1987 February 24/5 (Menzies *et al.* 1988; Catchpole *et al.* 1987; Menzies 1987). The first few spectra are in Fig. 40 (see also Plate 6).

The first spectrum showed a blue star (there is lots of light at the ultraviolet and blue end of the spectrum, but less light at the red end) with very broad, shallow absorption lines. The deepest absorption line appeared near 6200 Å, extending between 5900 and 6400 Å. It was this spectral line which Menzies described to Warner over the telephone which made them conclude that SN 1987A was Type I: they interpreted it as the silicon line at 6150 Å, which is very characteristic of a Type I supernova, and this was the conclusion which appeared in the second IAU astrogram. It was in fact the absorption dip from the H-alpha spectral line, which is normally at 6563 Å. The most extreme expansion velocity, represented by the blue extremity of the absorption dip at 5900 Å, was 30 000 km/s (0.1 the speed of light). The average speed of the material causing the absorption dip was at 18 500 km/s.

Over the next week the hydrogen lines became clearer and the emission peaks in the P Cygni profiles much more prominent. The blue light from the supernova faded rapidly. The velocity of the absorption dip of the H-alpha line

Fig. 40. Optical spectra of the supernova cover the first two months of outburst (dates of the month on the right hand side). At 6563 Å a vertical line marks the rest wavelength of H-alpha, the principal Balmer line of hydrogen. It shows as an S-shaped signature of absorption and emission in the spectra, the absorption dip moving from a position some 300 Å bluewards of 6563 Å on February 24 to 150 Å bluewards on April 6. H-beta at 4861 Å shows the same signature in the early spectra but is confused with other features near 5000 Å in later spectra. These and the features near 6000 Å are complicated blends of spectral lines which develop as the supernova cools. Data from the SAAO (Menzies *et al.* 1988).

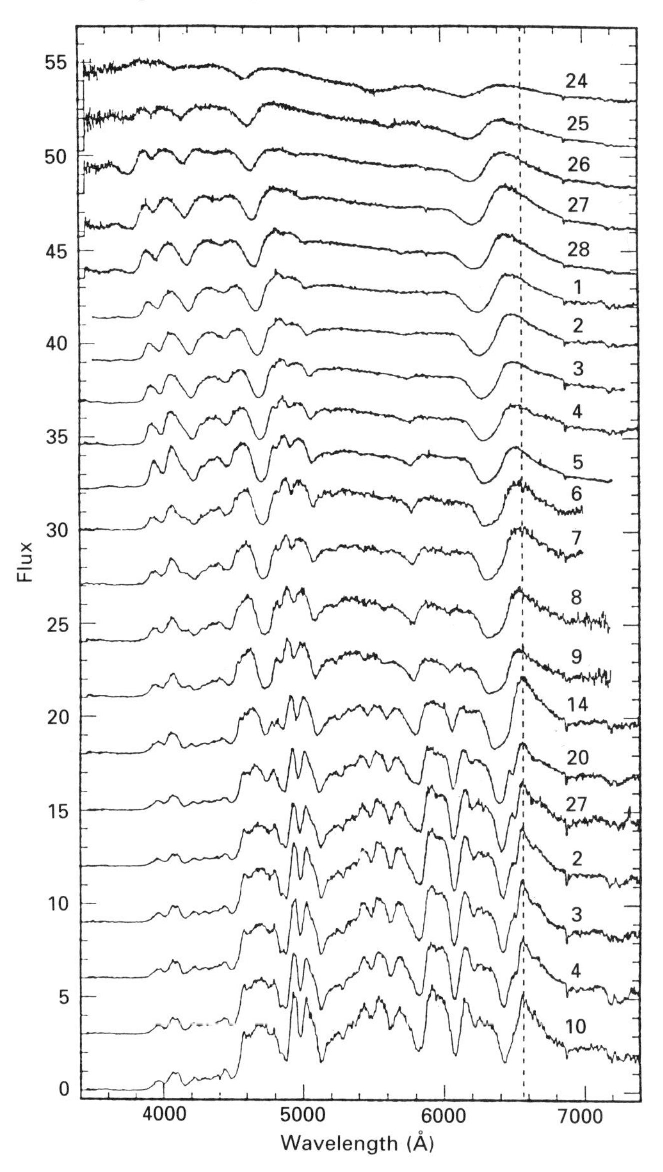

decelerated by an average of 780 km/s each day. Forty days after the explosion, it had fallen to 10 000 km/s. Although the explosion had started off rapidly, it looked as if it was slowing down; this, however, was an illusion.

As the supernova expanded the outer layers became less dense. Thus it was possible, from the outside, to see deeper and deeper into the supernova. Our line of sight penetrated deeper into the body of the star, seeking out the layers of the star which lie within the outer atmosphere. These inner layers were the slower moving layers which were following the outermost, faster moving layers. When they became visible, they showed a P Cygni profile which indicated that they were indeed moving slowly. It looked as if the supernova was slowing down but what was happening was that the slower bits were being revealed.

The supernova's brightness

The brightness of the supernova was measured by astronomers using ground-based and space-based instruments alike. The IUE satellite contains so-called Fine Error Sensors (FES) which are detectors which sense the positions of

Fig. 41. The dip in the H-alpha spectral line moved back towards the rest wavelength of H-alpha (6563 Å) as the supernova explosion progressed. From 1987 March 12 to 1987 May 27 the blue shift decreased from 230 Å to 100 Å, representing an apparent slow down of the explosion from 10 500 km/s to 4600 km/s. In fact this represents a change in the layer of the supernova which lies above the optically thick zone from a high, fast moving layer to a deeper, slower moving layer. Diagram from the AAT.

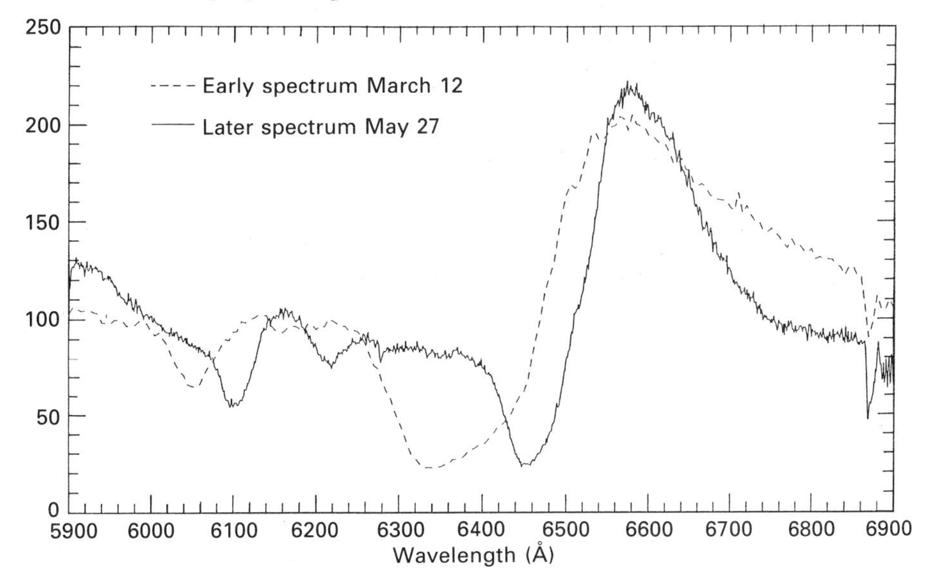

stars which happen to lie near to the star of interest and adjust the position of the satellite so that it can point accurately to this star. (Ground-based astronomers call this 'autoguiding', and they call FESs 'guide probes'; recall that Shelton wondered, when he first saw the supernova's image on his discovery photograph, why he had not used the supernova as a guide star to track his telescope.) The FESs can be made to point to the programme star, if it is bright enough, and can be used to record its magnitude. A fine series of measurements was produced by the satellite throughout its observations (Plate 5).

Amateur astronomers, principally in Australasia but also in Southern Africa and South America, also produced a creditable series of observations of the supernova throughout its visibility, tracking its brightness by eye and comparing it to other stars nearby to determine how bright it was. Figure 42 compiles data listed in the IAU Circulars and represents the observations of: T. Beresford† of Adelaide, South Australia, J. da S. Campos of Durban, South

† As a measure of Beresford's dedication, he observed the supernova not only often, but in particular on both Christmas night and New Year's Eve, 1987.

Fig. 42. After a rapid rise to magnitude 4.0 and a dip to 4.3, the supernova light curve peaked in 1987 May at magnitude 2.8 in this plot based upon amateurs' eyeball observations. From 1987 July, the supernova declined at 0.01 mag per day, dropping below naked eye visibility at the end of the year.

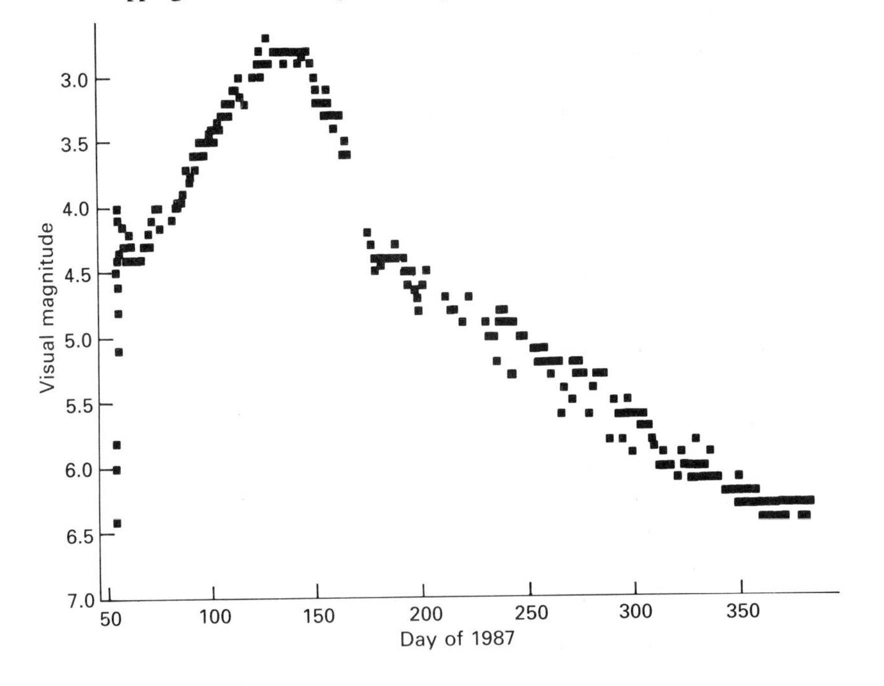

Africa, G. Garradd of Tamworth, NSW, C. Henshaw of Zimbabwe, R. McNaught of Coonabarabran, NSW, M. Morel of Rankin Park, NSW, V.F. de Assis Neto of Sao Francisco de Oliveira, Brazil, D.A.J. Seargent of The Entrance, NSW, and G.W. Wolf of Wellington, New Zealand.

The amateurs' observations show the sudden spectacular rise in brightness of the supernova on 1987 February 23–4. According to the observations by eye, the supernova reached a maximum brightness of magnitude 4.2 about February 25. The supernova declined in brightness by a few tenths of a magnitude and then levelled at magnitude 4.4 for about eight days. It then brightened steadily to reaching a maximum at magnitude 2.8 about May 20. It fell sharply back to magnitude 4.4 on July 1 and then began a slower, steady decline in magnitude which it followed into 1988.

Measuring time in astronomy

Because there are different numbers of days in the months of the year, and, in addition, the number of days in leap years is different from the usual 365, it is not straightforward to relate one day to another: to determine the time which has elapsed between, say, 1987 February 23 and 1988 November 11 requires a detailed calculation which it is easy to get wrong. Astronomers have set up a calendar which simply numbers the days one after another. The number in the sequence of dates is called the Julian Day, and the count started from zero at noon UT on January 1, 4713 BC. The neutrino burst from SN 1987A was received at Earth on JD 2446849.816458. Some graphs about the supernova which are reproduced in this book are expressed in terms of the calendar date, some in Julian Day Numbers and some in terms of the number of days which have elapsed since the explosion. Appendix 1 relates these methods of counting time.

Photometry

Impressive though the amateur measurements are, they give an incomplete picture of the brightness changes of the supernova. The measurements were made with a particular detector – the human eye. The measurements with the IUE FESs were made with the photomultipliers on the satellite. Both these detectors are responsive to a particular band of wavelengths and cannot see outside a limited spectral region – into the infrared and the ultraviolet, for instance. On the other hand, within their band of sensitivity they muddle up light from all wavelengths into a single measurement.

For the most meaningful work it is better first to split the radiation from the

Fig. 43. The change in magnitude of the supernova over the first year is recorded in a montage of photographs from the UK Schmidt Telescope © ROE.

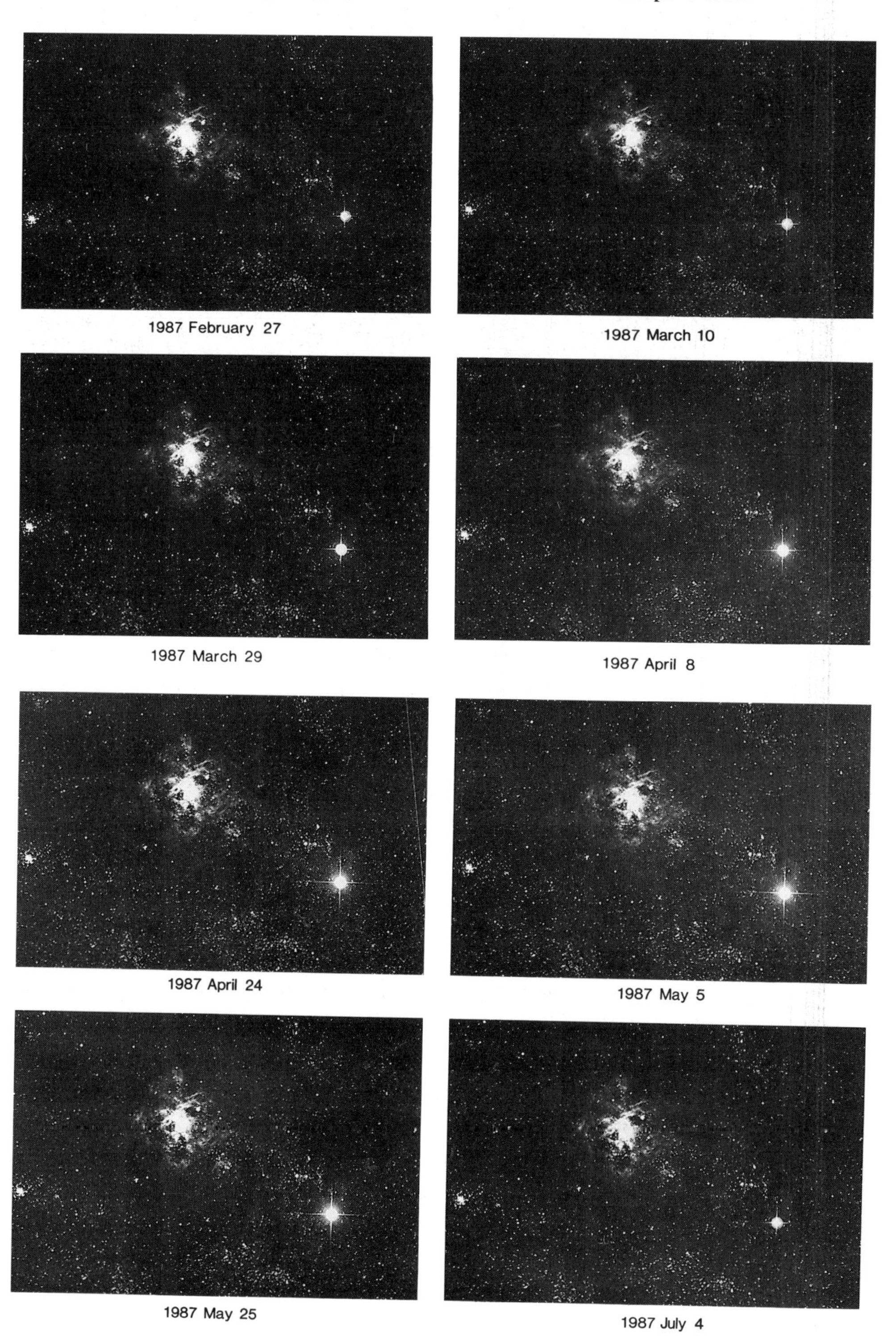

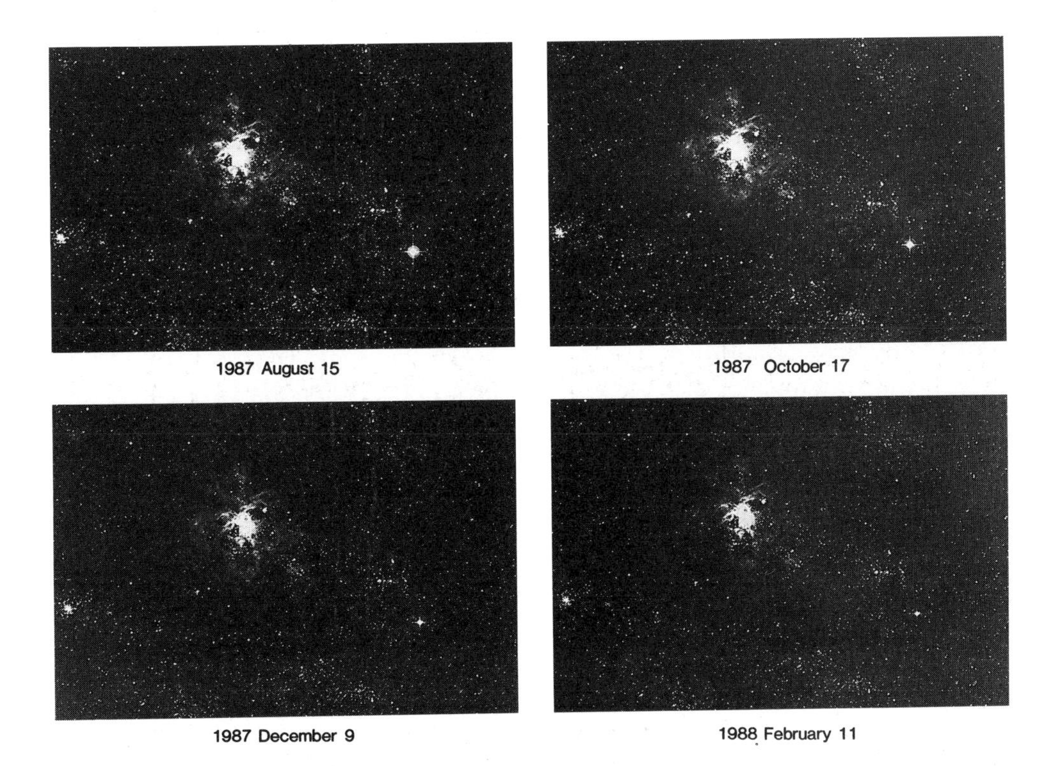
1987 August 15 — 1987 October 17 — 1987 December 9 — 1988 February 11

supernova into particular, well-defined wavebands, to make accurate observations, and then to combine the measurements in some way which represents more truly the physics of the supernova process, rather than accidents of biology or technology. As it happened, for most of its first eighteen months SN 1987A radiated in the ultraviolet–optical–infrared region; these wavelengths pass relatively easily through the Earth's atmosphere and the supernova's spectrum was thus accessible from the ground. To give the most useful total picture of the brightness of the supernova, professional ground-based astronomers measured the brightness of the supernova practically daily in several passbands.

Accurate, continuous series of observations were made particularly by the SAAO (Menzies *et al.* 1988, Catchpole *et al.* 1987) and by the CTIO (Hamuy *et al.* 1988). The SAAO daily observations covered the wavelength range from the ultraviolet (wavelength 3600 Å) to the near-infrared (3.5 microns); infrared observations were sometimes extended to 20 micron wavelengths. The measurements were made with coloured glass filters covering the detectors, to isolate the various wavebands which are traditionally known by letters of the alphabet (Table 20).

Table 20. *Photometric wavebands*

Letter	Name	Average wavelength (microns)
U	ultraviolet	0.36
B	blue	0.44
V	visual	0.55
R	red	0.70
I	near-infrared	0.90
J	infrared	1.25
K	infrared	2.2
L	infrared	3.4
M	infrared	4.8

Fig. 44. The magnitude of the supernova showed different behaviour depending on the waveband in which it was observed. In ultraviolet (U) and blue bands (B) it rapidly faded at first, but in redder light and in the infrared it brightened from the start, peaking about 90 days after outburst. Data from SAAO.

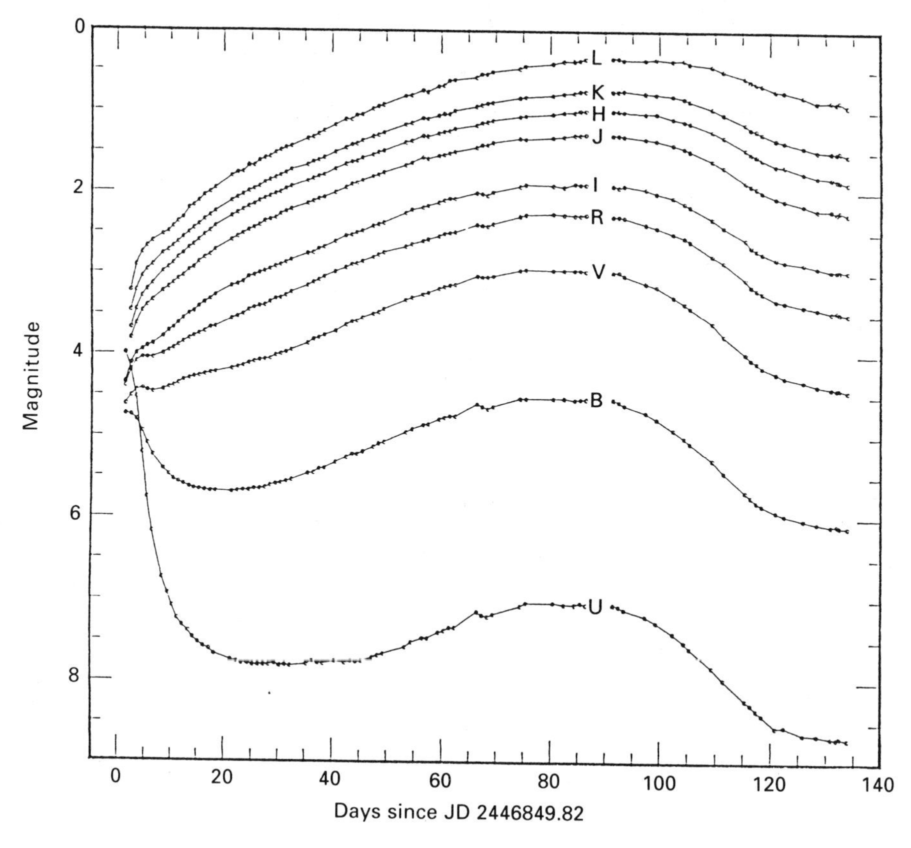

Figure 44 (Catchpole *et al.* 1987) shows the brightness of the supernova for the first 140 days. Clearly there were differences in the way the supernova seemed to behave, depending on what waveband it was observed in. In the infrared wavebands it progressively brightened right from the beginning, reaching a maximum brightness three months after Shelton's initial discovery. In the ultraviolet, on the other hand, the supernova's brightness at first fell dramatically, levelling off near magnitude 8 at the end of March, before rising slowly to a maximum in the middle of May. In the blue, too, it faded initially (but not so dramatically) but then began to brighten again from March 12, passing to a steeper maximum in May. In the V-, R- and I-bands the supernova brightened at first, faded a little and then brightened consistently to mid May.

In all wavebands the supernova faded quite rapidly from mid-May to the end of June, and then its rate of fading slowed quite abruptly, beginning a steady decline from about July 1.

Figure 45 shows the first ten days of the supernova's brightness as measured with amateur equipment (mostly by eye) and by the SAAO photometers, taking

Fig. 45. Amateurs' naked eye and photographic observations (open squares) in the first days of the supernova show a slight overshoot to magnitude 4.1 or so, compared to the later V-band photoelectric measurements (filled squares), which remain constant at magnitude 4.5 over this interval. But this is due to the blue sensitivity of the unfiltered eyeball measurements at a time when the supernova was markedly more blue than it became after only two days.

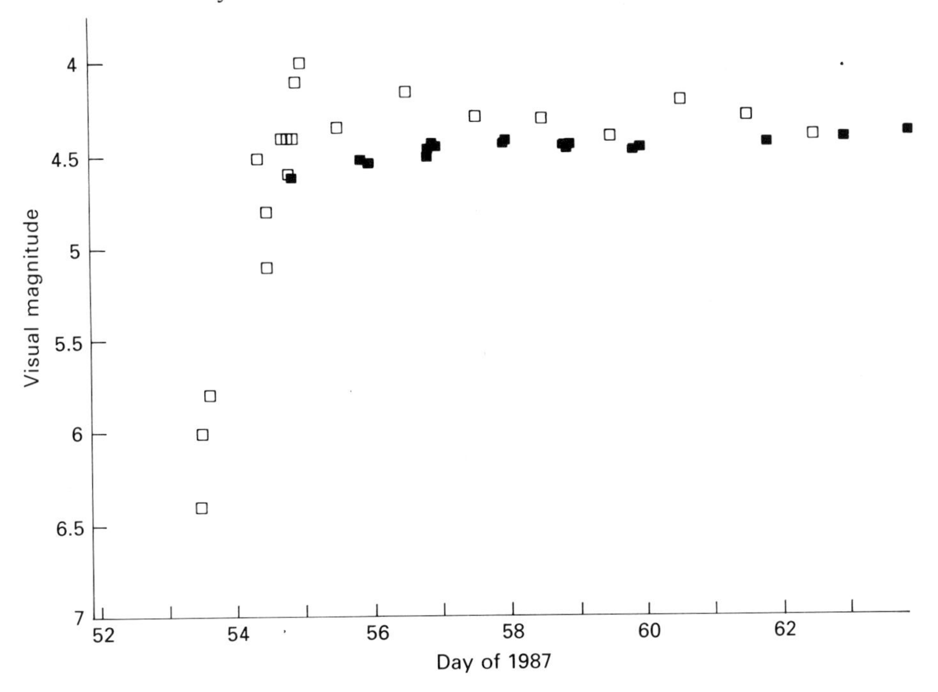

the V-waveband as most representative of what is seen by eye. The major difference between the photometric measurements and the eye measurements is that the supernova was observed by eye to be somewhat brighter between February 24–7 than by photometry. This is because the eye is sensitive to blue light excluded by the glass filter from the photometric V-band measurement and the supernova was radiating considerable blue light at that time. Once the blue of the supernova had faded away, the eye and the V-band measurements were in good agreement.

Colour and temperature

The ratio of the blue light to the red light from a star is regarded by astronomers as a measurement of the colour of the star; in magnitudes the colour is called B–V. The supernova had B–V = 0.12 on February 24, which is a neutral colour – neither blue nor red. But the supernova very rapidly became red, as B–V increased to 1.6 by the end of March. This numerical fact corresponded with the naked eye appearance of the supernova, as astronomers watched it turn from white to red.

The change in colour can be seen on the colour pictures taken by numerous astronomers in the first months of the supernova (Plates 1, 4).

The brightness of the supernova in the various passbands of Table 20 can be plotted as a function of wavelength to show the overall energy distribution of the radiation from the supernova. This can then be fitted with the spectrum expected from a black body and the temperature determined corresponding to

Fig. 46. The photometry of the supernova at the several wavebands from ultraviolet to infrared can be used to calculate the overall spectrum of the supernova (filled squares). Although the spectra are quite complex, with bumps and dips, black body curves can be made to give a reasonable overall fit and determine the temperature of the supernova. SAAO data.

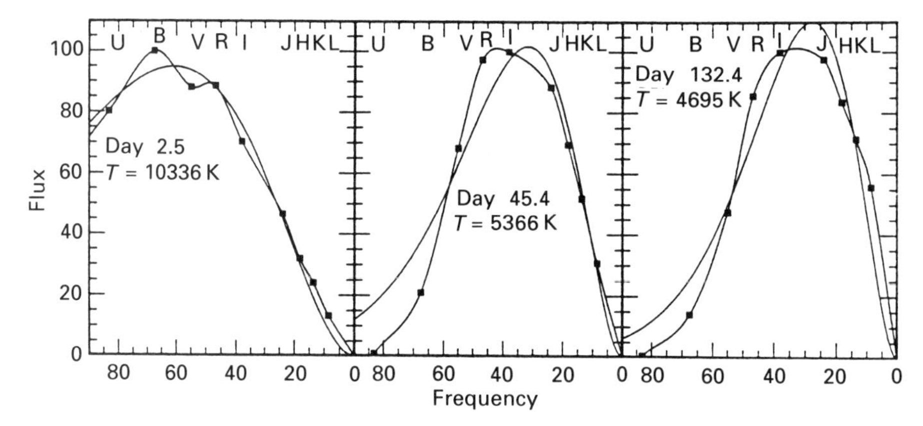

the black body which fits best. Considering the complicated shape of the spectrum of the supernova, the black body fits are quite good. The temperature of the supernova declined rapidly from 14 000 K on February 24 to 6000 K on March 5; it settled at 5600 K for three months, somewhat cooler than the Sun, and then dropped early in July to just below 5000 K.

Once the overall energy distribution has been established it is possible to calculate the total energy flux from the supernova in all radiated bands as a function of time. Since the supernova is in the LMC, whose distance is known to good precision (about 10%), it is possible to calculate the energy flux fairly precisely, the only uncertainty being in the correction for the small amount of radiation absorbed in the passage of the radiation through interstellar dust.

The supernova very rapidly faded, halving its total energy output from 4×10^{41} erg/s in four days from February 24 to February 28. This corresponds to the rapid cooling of the supernova in that initial period. SN 1987A faded five times faster than other Type II supernovae. At first astronomers were very puzzled as to why this should be.

Fig. 47. From the fits of black body curves to the photometry, the temperature of the supernova can be determined. The temperature plummeted in a few days to 5500 K and held that value for several months before dropping again to 4500 K. SAAO data.

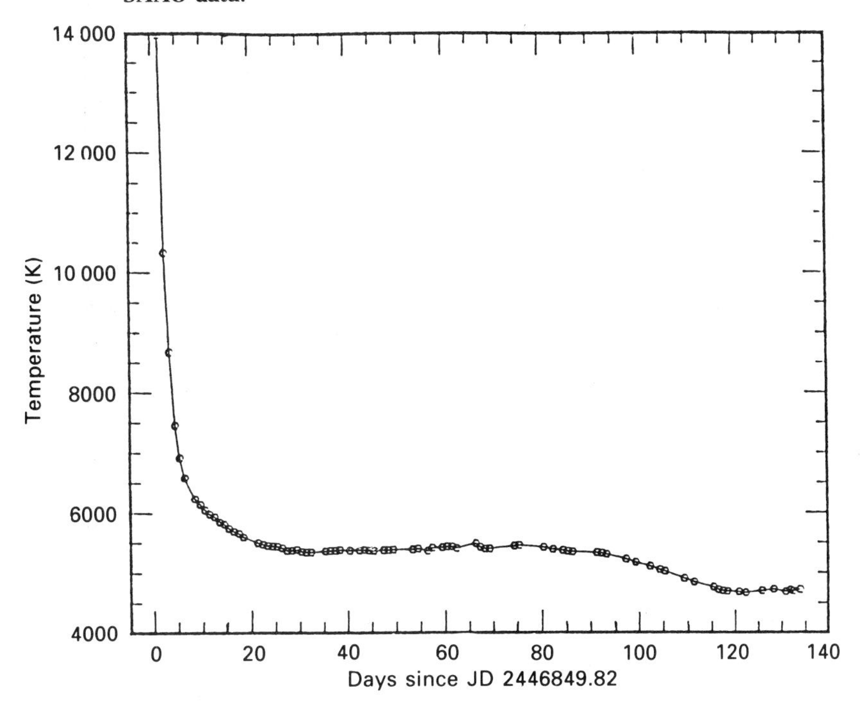

But from March 1 the supernova increased its total energy output by a factor of 4 from 2×10^{41} erg/s to 8×10^{41} erg/s at its maximum at the end of May, fading rapidly from that time. By July 1, the supernova had declined to 3×10^{41} erg/s and then began to fade very regularly. In fact, when the total luminosity of the supernova is plotted logarithmically as a function of time it shows a straight line whose slopc is practically unchanged throughout the remainder of 1987. This straight line was a crucial clue to SN 1987A's peculiar behaviour, and direct evidence of the creation of the chemical elements in stars (see Chapter 10).

The size of SN 1987A

The photometry of SN 1987A gave fundamental data on the luminosity and temperature of the supernova with an unprecedented accuracy and completeness. These fundamental data make it possible to plot the size of the supernova from the following argument. (Size refers to the layers of the

Fig. 48. The black body fits to the photometry also determine the total flux of energy which radiates from the supernova. In spite of the wider spectral information which goes into these data their plot against time is still called a 'light curve'. The light curve of SN 1987A had a shape which has not been seen in a supernova before, with a very rapid initial drop, and then a steep climb to a broad maximum, followed by a slower decline. SAAO data.

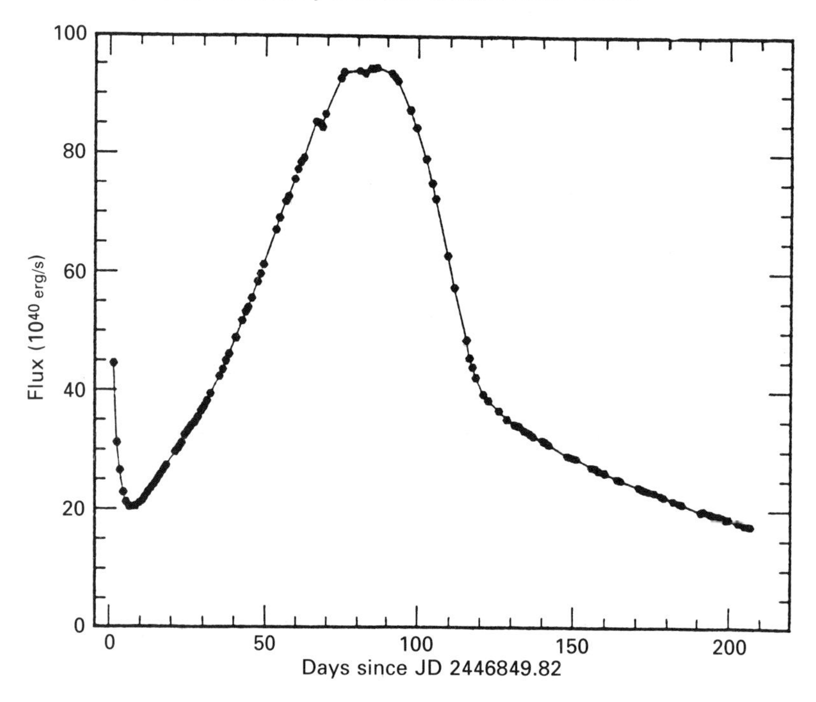

supernova emitting most of the light and infrared which we see at any one time; the transparent outer parts would be much bigger.)

The luminosity L and the temperature T of a sphere are connected with its radius by the following relation known as the Stefan–Boltzmann law.

$$L = 4\pi R^2 \sigma T^4.$$

Fig. 49. Ultraviolet measurements by satellite, and visual and infrared measurements by ground-based telescopes, have been combined over the first 60 days of the supernova to show how the supernova cooled rapidly as it expanded: the peaks of the black body curves moved towards the low frequency infrared radiation, but then the supernova brightened without changing temperature. Astronomers were initially very puzzled by this behaviour.

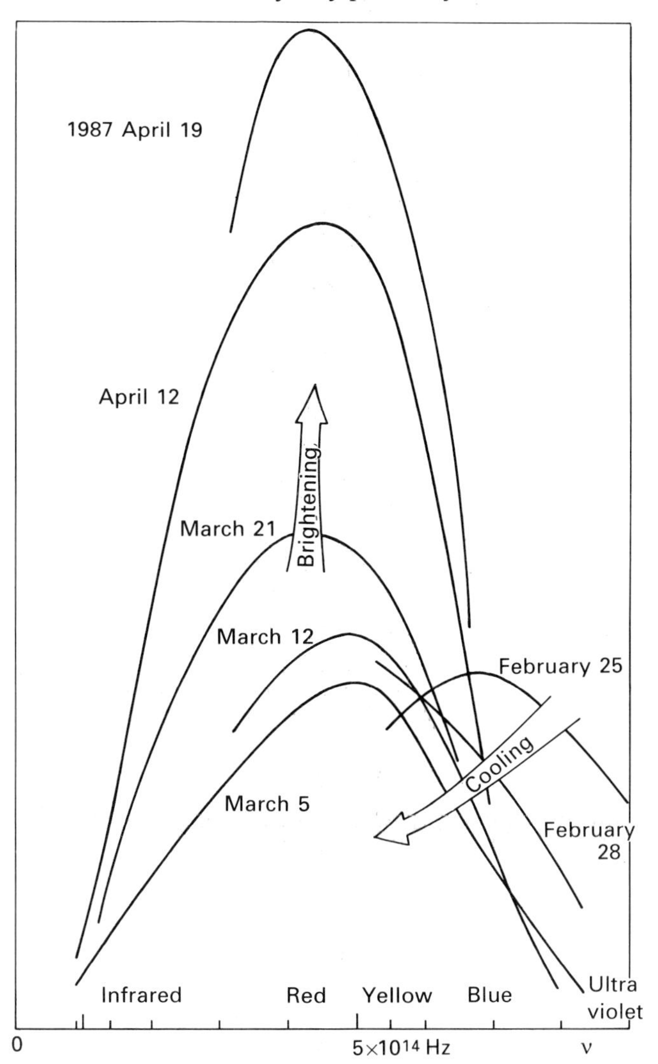

Here $4\pi R^2$ is the area of the surface of the sphere, and σ is a physical constant called Stefan's constant:

$$\sigma = 5.7 \times 10^{-5} \text{ erg/(cm}^4 \text{ s}^1 \text{ K}^4).$$

Sk − 69 202 at the moment of core collapse had a radius about 2×10^{12} cm (70 light-seconds – $\frac{1}{3}$ the radius of the orbit of Mercury). SN 1987A reached its maximum size 100 days after the outburst of 1.4×10^{15} cm radius (0.5 light-days – the radius of the orbit of Saturn). After that time, SN 1987A had expanded and cooled so much that its apparent size began to decrease, as we began to see deeper and deeper towards the centre of the expanding envelope (although the explosion was still speeding on).

Fig. 50. The radius of the supernova, as determined by the size of the photosphere from which the light and infrared radiation came, increased, steeply at first but then more slowly, reaching a maximum about 100 days after the outburst. Then the supernova appeared to begin to decrease in size as the supernova became more transparent and the photosphere began to drop rapidly back into the expanding shell. SAAO data.

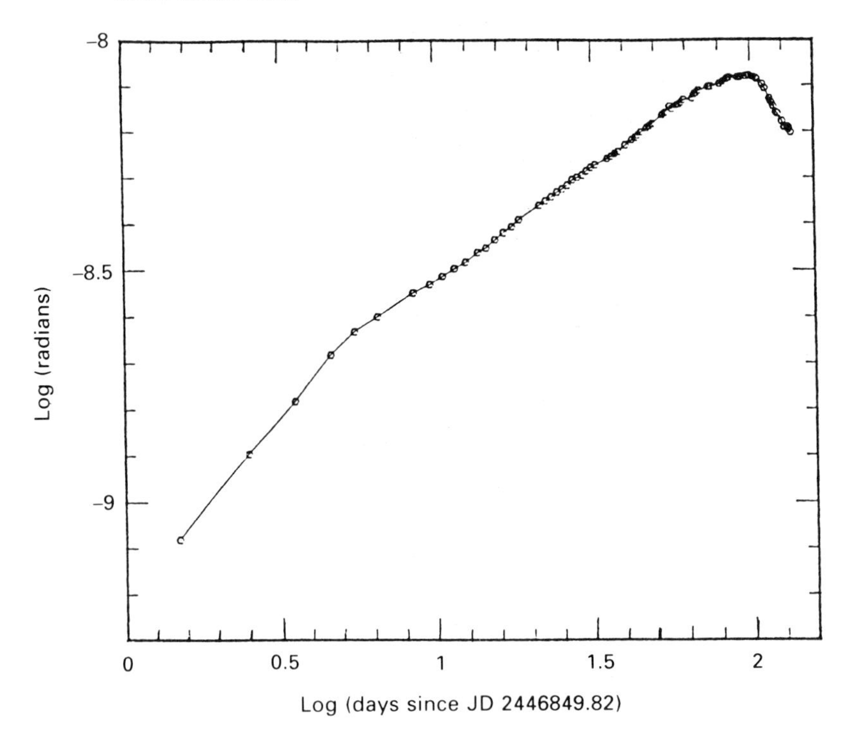

Radio emission from SN 1987A

Radio emission produced by the rapidly expanding supernova was detected from SN 1987A on the day after its optical discovery. First to see the supernova were the two radio telescopes operated by the University of Sydney in Australia, the Molongolo Observatory Synthesis Telescope (it's the MOST) which observes at a frequency of 843 MHz and the Fleurs Synthesis Telescope at 1.4 GHz. Both telescopes saw the supernova when it rose above the horizon in the afternoon of February 25. The Parkes–Tidbinbilla interferometer (2.3 and 8.4 GHz) was brought into operation to observe the supernova on February 26.

In contrast to the optical emission from the supernova, which far outshone any other object in the LMC, SN 1987A was a relatively weak radio star amongst the neighbouring, much more luminous nebulae and supernova remnants (Fig. 7, p. 24).

At the lower frequencies the radio intensity rose rapidly and peaked on February 26 and 27. At the higher frequencies the radio intensity was already falling when the supernova was first observed on the 26th. The supernova faded away rapidly, and became invisible to the three higher frequencies by March 1; the MOST continued to detect it, however, as a very faint radio source, long afterwards (Turtle *et al.* 1987; Manchester 1987; Bartell *et al.* 1988).

The radio pulse from SN 1987A was radiated from a 'radiosphere' due to electrons accelerated at the outer shell of the material thrown out from the supernova, being driven ahead of the shell as if it was a piston. The piston caught up and compressed the magnetic field of the star's surface; the electrons were accelerated by the shock wave as it broke out from the surface. The electrons moved back and forth, trapped in and spiralling around the magnetic field. They emitted synchrotron radiation. The magnetic field of the star diminished rapidly as the radiosphere expanded, and the synchrotron radiation quickly faded away as the supernova grew bigger.

The supernova was observed not only from Australia but also, in a coordinated observation, from South Africa. The 34 m Tidbinbilla dish and the 26 m dish at Hartebeesthoek jointly observed the supernova on February 28, March 1 and 2 by the technique of Very Long Baseline Interferometry (VLBI). In some respects, the two radio telescopes together made the equivalent of a radio telescope the size of the Indian Ocean lying between them. The diameter of the equivalent radio telescope was 9400 km, the equivalent of 72 million wavelengths of the radio waves at which the two telescopes were observing. This giant telescope was able to determine that the supernova was not at that time an unresolved radio star but had structure and dimension. It was not possible to say what the size of the radiosphere was but only to give a minimum size, which when the supernova was aged 5.2 days was $> 8.3 \times 10^{14}$ cm. The expansion speed of the radiosphere was thus at least 20 000 km/s – this is consistent with the expansion speeds of the surface from the optical spectra and P Cygni profiles.

Was the supernova round?

The explosion of a supernova is nearly always represented in diagrams as a sphere – completely and uniformly round. Is this supernova really round?

Astronomers are faced with interpreting data which is evidence of situations which are hard to visualise; and they start always with the simplest idea. The simplest visualisation of a supernova is that it is round and the material of the star flows out uniformly and radially from the centre. This idea got astronomers started on the right lines.

But there is no reason why the explosions of supernovae must be precisely round. The fact that no-one has persistently questioned whether they are round means only that equations which are easier to handle because they assume spherical symmetry have not been critically tested by real supernovae viewed from a distance of 100 million light-years, from which everything looks simple. Because SN 1987A was so near and so bright there was a chance to use a technique which would tell if it was round or not.

That technique was polarimetry – measuring how the light from the supernova was polarised.

Polarisation is a geometrical property of a light wave, and that is the fundamental reason why polarimetry can reveal the geometry of the object it comes from. The polarisation of a light wave is the direction in which it oscillates – up and down or, at right angles to this, from side to side. A light wave which oscillates at an intermediate angle is envisaged as a combination of light waves in the two orthogonal directions. Most light is a mixture of all polarisations; if one direction is preferred more than another, the light is said to be 'polarised'.

A polarimeter is a device which separates light of different polarisations – a pair of Polaroid spectacles is a crude polarimeter. The lenses in such spectacles let through light which oscillates up and down but absorb light which oscillates sideways. When sunlight, consisting of light waves of all polarisations, obliquely strikes a horizontal surface, the horizontally (sideways) polarised light is reflected much more than the vertically (up and down) polarised light. The reflected light from a surface can dazzle by its intensity and cause 'glare'. Since the glare is more horizontally oscillating than vertically oscillating, it is polarised light and it can be absorbed by the Polaroid spectacles.

If, while wearing Polaroid spectacles, you view a road surface towards the Sun, the glare is eliminated; turn the spectacles through a right angle (to the 'orthogonal' direction) and the glare returns. The spectacles are thus telling you, in effect, that the road surface is horizontal, revealing one of its geometrical properties. The technique of polarimetry consists fundamentally in comparing in orthogonal polarisations the light which is received from the object.

Now, radiation from inside the supernova passes through its atmosphere; the radiation encounters parts of the atmosphere, and is deflected at each encounter. The radiation staggers on a crooked path which wends gradually upwards and outwards. Eventually its journey towards us brings the radiation to the top of the atmosphere, from which, as light, it passes without further interaction to the Earth.

At the last interaction with the supernova's atmosphere one polarisation of the light might have been preferentially set on its way, so the light is polarised. For example it might be polarised circumferentially around the supernova. When we view the supernova from a great distance we cannot view the individual parts of the structure of the supernova – all the polarisations are mixed back into a single stellar image. If the supernova is round then the circumferential polarisations average out to nothing – there's no net polarisation of the supernova light (Fig. 51).

But if the supernova is not spherically symmetric – say, the explosion is elliptical, but the polarisations still run circumferentially – then, when the polarisations are mixed up by being viewed from a great distance, there would be an excess of the polarisations from the elongated sides of the supernova.

The polarisation effects depend on which bit of the atmosphere created the light; and the P Cygni profile relates wavelength to atmospheric position. So the polarisation will depend on wavelength – if the supernova isn't round.

This effect was found in May at the AAO by Cropper *et al.* (1987). They have available a Pockels cell polarimeter which fits over the spectrograph attached to the telescope. The Pockels cell is an optical device which separates the light into two images consisting of orthogonal polarisations. The spectra of the two images can be ratioed, and if part of one spectrum dominates the corresponding part of the other, then a feature in the spectrum is polarised. At first, four days after the discovery of the supernova by Shelton, no polarisation features were

Fig. 51.(*a*) *A* cross-section of a supernova showing a photosphere surrounded by an atmosphere. Light from the photosphere is radiated and is scattered or deflected in a slice of the atmosphere, and polarised. It is turned towards the Earth.
(*b*) Two possible views of a supernova as seen from Earth; in one case the supernova is round and in the other case it is elongated. The polarisation of the scattered light is circumferential. In the symmetrical case, the up-and-down and the side-to-side polarisations balance and there is no net polarisation. But in the elliptical case, there is an excess of up-and-down polarisations because that is the direction in which the supernova is elongated. Thus a residual polarisation remains.

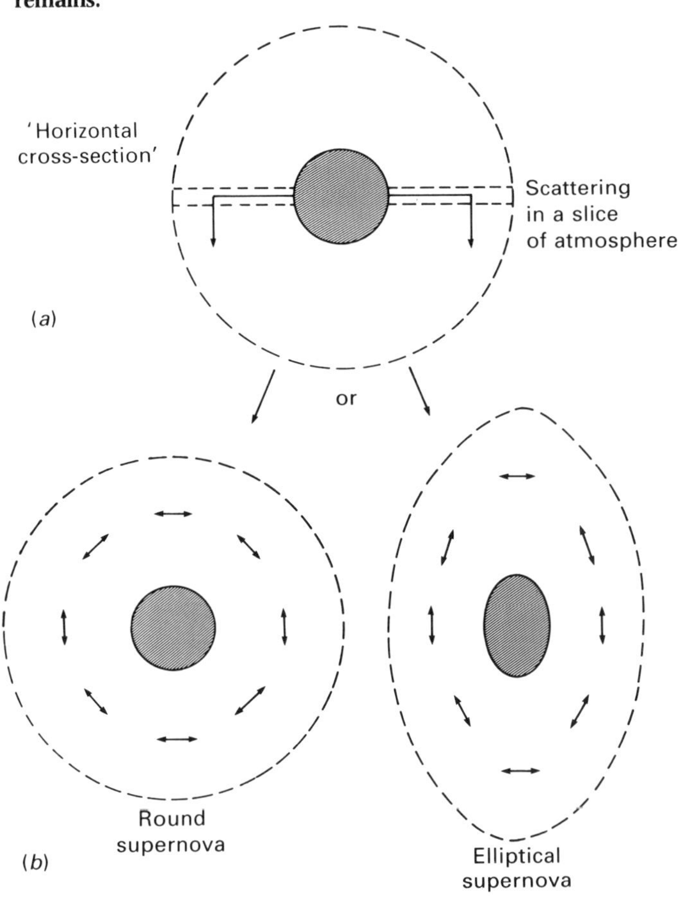

found. But some hints of polarisation changes were found in March and by May the degree and the direction of the polarisation of the light from the supernova was found to change markedly with the wavelength of the light. In particular, the polarisation changed across the P Cygni profile of the hydrogen emission lines.

Because the polarisation features of the supernova's spectrum changed with time they must have been intrinsic to the supernova, rather than some external feature of the passage of the light through the interstellar medium, which can also produce polarisation.

The amount by which the supernova is non-spherical is not marked. There are as yet only simple theories to interpret the observations, but they imply that the ratio of the axes of the ellipsoidal supernova is 0.8 to 0.9 – in other words, the

Fig. 52.(*a*) A spectrum of the H-alpha spectral line of the supernova (compare with Figure 41). The polarisation of the light in the spectrum is graphed in (*b*), and reaches a value of 2% in the tip of the H-alpha line; it varies over the rest of the profile of the spectral line. In (*c*) the direction of the polarisation is shown. It, too, changes dramatically with wavelength. Data from the AAO.

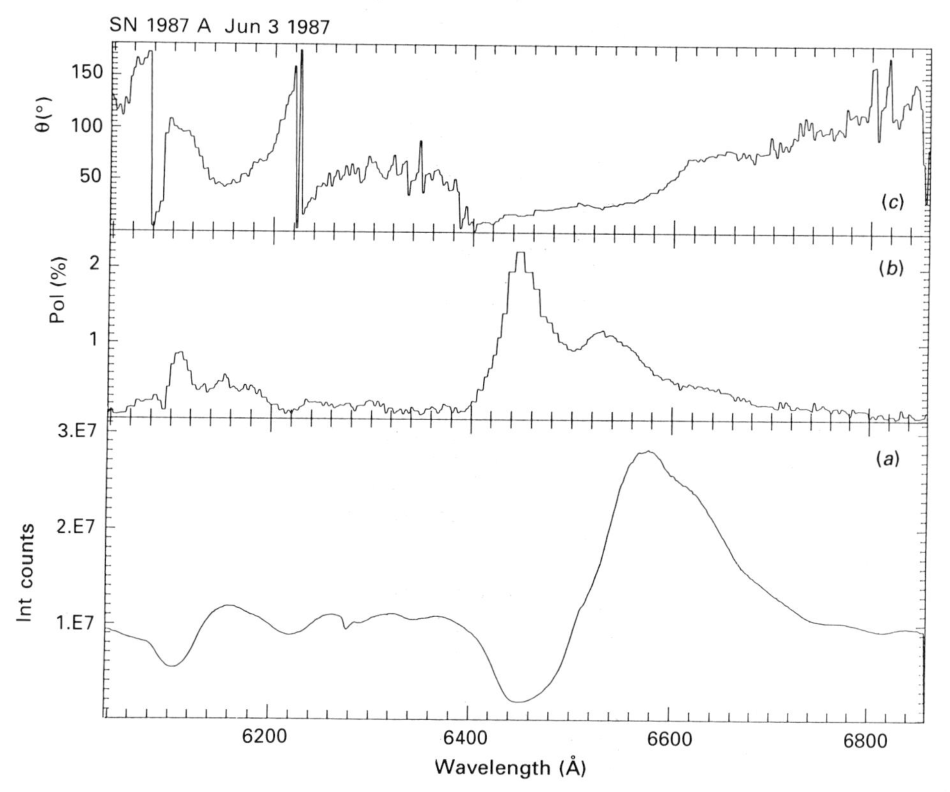

short axis is 10–20% shorter than the long axis. This is the same shape as a face which is 10–20% thinner from ear to ear than it is long from chin to crown.

In May 1988, direct evidence of the asymmetry of the supernova was discovered by the technique of speckle interferometry. Using this technique (see Chapter 11) images of the supernova were made, overcoming the blurriness of the Earth's atmosphere. The supernova could be seen to be elliptical (Plate 8).

Why should a supernova not be round? Astronomers have thoughts about this question but no very specific proposals. If the core is not precisely centrally placed in the star, due perhaps to the influence of a companion, then the detonation which starts the outflow of the supernova will be off-centre and might throw material preferentially in one direction and not in another. If the star was rotating, or had a strong magnetic field, then the expansion might be preferentially channelled along the axes. If the collapse was not radially inwards, but if the core 'pancaked', preferentially falling faster in one direction than another, then the release of energy and the explosion might not be symmetrical. The consequences of the asymmetric pancaking of the core collapse might be a very asymmetric neutron star – a binary star, possibly. Stella and Treves (1987) have sketched a 'collapse, pursuit and plunge' scenario which features a short-lived neutron binary star in SN 1987A (see Chapter 12).

These are conjectural explanations and no-one knows why SN 1987A is not round, nor whether this is generally the case for supernovae. However, the most recent known Type II supernova in our Galaxy, the one which occurred in 1054, left a nebula, the Crab, which is very clearly not spherical. Its axial ratio is 0.7. The Crab Nebula is evidence for the asymmetric explosion of the supernova of 1054.

Was the outflow uniform?

Another idealisation of the outflow of the supernova is that it is uniform – smoothly flowing like a wide, deep river, rather than turbulent like narrow, shallow rapids. It makes a difference which you assume when you come to calculate the visibility of the interior of the supernova, amongst other things. If the outflow is turbulent then it may be possible to view parts of the interior at an earlier stage than if the outflow was smooth. In the same way, a river bed may be glimpsed sporadically through the water as it splashes over rapids.

One piece of evidence about the smoothness of the outflow of the supernova came from very high resolution observations of the spectra of optical emission lines. AAT observations (Allen 1988) of the tops of the emission lines show wiggles which represent material moving in lumps, or fingers, at different speeds. The wiggles must represent something serious in the supernova, because they are identical on two separate spectral lines which arise from precisely the same material (the oxygen spectral lines [OI]at 6300 and 6363 Å).

No-one has seen this kind of fine detail in a supernova before and theories to describe it are new (Nagasawa, Nakamura and Miyama 1988 – see Plate 7).

The gross properties of the spectra of supernovae, on the other hand, have been known for decades, and there are quite well-developed theories to model them. These theories have been tested by the rather unusual circumstances of the LMC supernova.

Modelling the spectra

The idea of calculating a synthetic spectrum is to use theory to represent the atmosphere of a supernova. The technique can be very complex in its details, but its aim is to test what simple assumptions about the atmosphere are necessary to reproduce what astronomers observe. The reason why it is necessary to do the whole calculation is that, with prominent exceptions like

Fig. 53. Small scale fluctuations in the oxygen lines of the supernova match in both lines, confirming that the structure is intrinsic to the supernova; perhaps the bumps represent 'fingers' of material moving outwards from the supernova at the speeds represented by the Doppler shift of the bumps. Data from AAO.

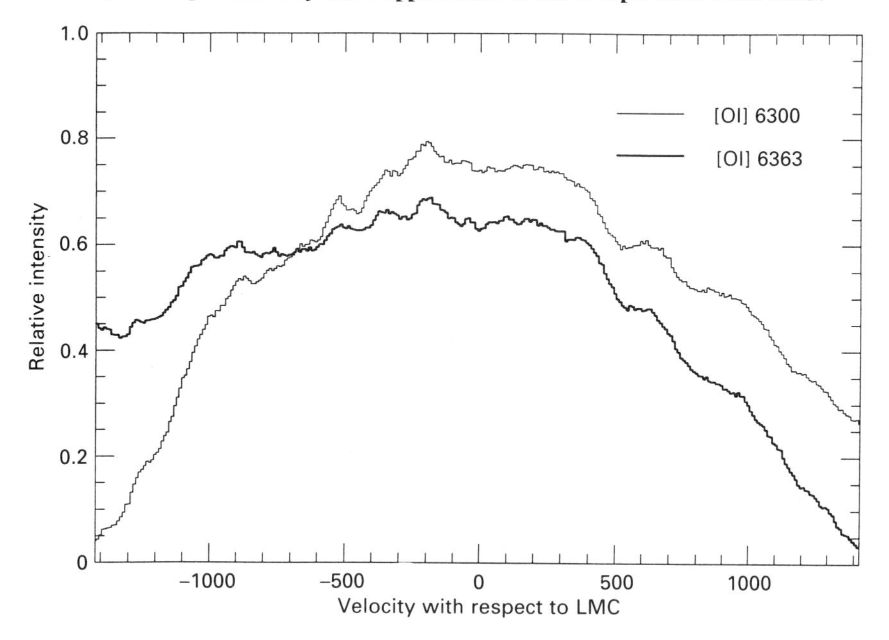

the big emission lines, the dips and bumps do not correspond to specific spectral lines or even to atoms or ions. They are blends of features and their individual components are hard to guess.

As a gross example consider the spectrum of the supernova in the ultraviolet. When the spectrum was first recorded there was a considerable amount of ultraviolet radiation, and the IUE satellite was able to obtain very good records of the supernova. In the first week the ultraviolet flux dropped very rapidly. Its disappearance was due to the effect called 'line-blanketing' (Plate 6).

Numerous atoms and ions have many spectral lines in the ultraviolet region. There are so many that when they exist in the atmosphere of a star they absorb all the ultraviolet radiation (the energy reappears elsewhere in the optical region, as the ions re-radiate it). The disappearance of the ultraviolet cannot be attributed, however, to any individual ions but in general to all the heavier elements. Synthetic spectra are intended to calculate this kind of effect – and, as Leon Lucy's work at ESO shows (Lucy 1987), can do so with striking success.

The complication in constructing synthetic spectra lies not in the astronomy of the supernova but in the physics of the atoms which contribute to the appearance of the spectra. There are 100 atoms in the periodic table – fortunately not all of them are from abundant elements, but even if the least abundant are eliminated from consideration, dozens are left. Each atom has several electrons which can make numerous transitions which give rise to spectral lines. Each atom can appear as several ions (neutral atoms, atoms with one electron missing, two electrons missing etc.). In the most complicated synthetic spectra made of SN 1987A, 200 000 spectral lines are tabulated in a computer to form the database of atomic physics to work with.

Once the right kind of database of spectral lines and, of course, the correct physics equations to handle them have been programmed, astronomers have to put the astronomy into the analysis. What sorts of assumptions about the supernova need to be made? – what is the astronomy which is being tested?

The abundances of the different elements form a key set of assumptions. The abundances are expressed in terms of the numbers of atoms of various elements relative to the most common element in the universe, hydrogen. How many atoms of, say, helium are there in comparison to hydrogen? What is the ratio of iron, or other elements like barium, to hydrogen? The starting assumption would be that the supernova was like the Sun and had the elements in the same ratio as in the Sun. But then astronomers would want to take account of the fact that in nebulae in the LMC everything apart from helium appears to be one quarter or so less abundant than in the Sun. The assumption that the supernova is like everything else in the LMC might not be accurate, however – Sk – 69 202 has had its own history which may have altered its particular chemical abundances. For example the carbon and oxygen in the hydrogen burning layers may have been converted to nitrogen by the CNO cycle (see Chapter 5). Establishing the chemical history of Sk − 69 202 by the technique of synthetic spectra would be an interesting piece of the jigsaw. In the supernova itself elements may have been manufactured – astronomers will want to know if we can see evidence of this.

Varying the assumptions, theoreticians calculate numerous representations of the supernova's spectrum as different stages in its evolution. They match the calculations with what is observed and try to home in on key differences between calculation and observation, honing the assumptions until they get a good fit. The results seem to be these.

In its early stages, SN 1987A seems to have had elements at rather normal abundances, so far as can be told from the synthetic spectra (Branch 1987). It seems that there was somewhat more helium (two or three times more) than in the Sun (Lucy 1987). This makes astronomers think about how much mass Sk − 69 202 lost before it exploded and whether the layers below the surface of the star have been exposed. In these layers some hydrogen has been processed to helium. It may be that overall most of the elements other than helium were less abundant than in the Sun (Hauschildt, Spies, Wehrse and Shaviv 1987). This fits well with astronomers' knowledge of the LMC.

The temperature structure, size and speed of the outflowing supernova affect the calculations of the synthetic spectra. So does the way in which the density of the envelope decreases with radius.

This last effect gives a way to probe the structure of a star – not many measurements of stars tell much about the distribution of matter inside the surface layers. It turned out that the density in the supernova rose very steeply the further it was penetrated – the density was inversely proportional to the radius of the star to the tenth power:

$$d \propto r^{-10}.$$

This, in fact, is the reason that the line-blanketing of the ultraviolet appeared so quickly. It means that the star which exploded was rather compact – not at all the extended envelope which would be typical of a red supergiant – and red supergiants had been assumed to be the sites of supernova explosions until the appearance of SN 1987A. With hindsight we know exactly why there was this difference: Sk − 69 202 was a *blue* supergiant.

The construction of the synthetic spectra depends on the assumptions of spherical symmetry in the supernova: if it is not round, the spectra are different. There is some evidence from the synthetic spectra that SN 1987A is not round, consistent with the results from the spectropolarimetry.

The success of synthetic spectroscopy in dealing with SN 1987A was very encouraging: SN 1987A was a warm-up exercise for being able to treat supernovae in the future which have really funny abundances. But one residual puzzle from synthetic spectra calculations of SN 1987A lay in the barium problem.

Barium produces a spectral line near to H-alpha. A line at this position appeared very prominently from the end of 1987 March to the end of June. The conclusion was that barium was particularly strong. This was borne out by the calculations of synthetic spectra, which suggested it was five times overabundant compared with the Sun. This raised the question of whether barium had been made in the supernova explosion, and provoked speculation on how the barium got from inside the supernova to the visible outer parts so early. Maybe the strength of the barium line is a subtlety of the technique of constructing synthetic spectra which is not understood; alternatively, maybe the 'barium' line near H-alpha was not due to barium at all but was a fragment of fast moving hydrogen, producing an H-alpha line Doppler-shifted by coincidence to the barium wavelength. Astronomers who focused on this detail could not agree on its true significance.

Carbon monoxide

As the surface which we see sinks further into the deeper layers of the supernova we can expect to see the effects of the various zones of the 'onion' structure of which Sk − 69 202 was formed before it exploded. The first evidence of this appeared in June 1987 (day 110). It took the form of the detection in the infrared of molecules of carbon monoxide. Carbon monoxide emits bands of spectral lines at 4.6 microns (the M photometric region, see Table 20, p. 138) and 2.3 microns (the K region). As time went on, other bands appeared.

The carbon monoxide was shown to be coming from regions with speeds of about 1500 km/s. This was too large for the carbon monoxide to be located in the stellar wind of Sk − 69 202, left over from the time when it was a red supergiant. It must therefore be in the supernova mantle, in the region where the carbon and oxygen had been made by Sk − 69 202 (Danziger *et al.* 1988; Spyromilio 1988).

But not only did SN 1987A reveal, in the dismembered body of Sk − 69 202, material which the progenitor had made; it revealed, too, elements which had been made in the very detonation of the supernova itself.

10 Creation of the elements

Nucleosynthesis in supernovae

The core of Sk – 69 202 formed a neutron star, releasing floods of neutrinos, but that was not the only nuclear process which occurred in the supernova explosion. Temperatures in the core of the supernova explosion rose to 10^{10} K; this heat and the shock from the core collapse also heated the material which enveloped the core. The atomic nuclei in this material were formed by nuclear burning – neon, oxygen, carbon, helium. Because of the high temperatures, the nuclei were rapidly agitated, moving at high speeds. When the nuclei encountered one another they collided with considerable force, enough to merge them together to form new kinds of nuclei. This is the process called nucleosynthesis.

The material surrounding the core had a high density – it had a high density in Sk – 69 202 just before the explosion and the shock wave which propagated from the core compressed the material to even higher densities. This meant that the nuclei were packed closely together. Because the nuclei were close and travelling fast over a short distance they collided very often. Thus if a collision resulted in the formation of a big, *un*stable nucleus, which after a brief life was likely to decay, there remained the possibility that it would take part in a further collision after an even briefer time and build up into a bigger, *stable* nucleus. Only in extreme conditions of pressure, temperature and density can chains of nucleosynthesis occur which rely on the rapid addition of new particles to unstable nuclei. This means that the nuclei which are made by this process are rather rare, because the conditions under which they are manufactured are uncommon.

Table 21. *Nucleosynthesis from a 15 solar mass star*

Nuclide	Amount produced (solar masses)
^{4}He	5.3
^{12}C	0.2
^{14}N	0.02
^{16}O	0.02
^{20}Ne	0.024
^{24}Mg	0.005
^{28}Si	0.05
^{32}S	0.03
^{36}Ar	0.006
^{40}Ca	0.006
^{44}Ti	0.0002
^{56}Ni	0.12

The process of nucleosynthesis which takes place along this rapid path is called the r-process; supernovae are the main places in the Universe where it can occur. Thus supernovae are cosmic factories which build the nuclei of complex elements from the more simple nuclei. In the explosion, supernovae distribute what they have manufactured.

The main products in SN 1987A were the alpha-nuclides, which are the nuclei which are made up of clusters of four nucleons – two protons plus two neutrons. This is a particularly stable combination of nucleons – physicists describe this, in an anthropomorphic phrase, by saying that nuclei 'want' to be in clusters of four nucleons; of course, nuclei do not have will, nor do they have the ability to see the future and evaluate its benefit to them, and so it is not literally true that they can desire anything. But, as it were, left to themselves in a collection, and agitated with heat and pressure, nuclei arrange themselves in groups of four like the alpha particles, or helium nuclei.

The first thing that the exploding shock wave of SN 1987A saw as it set off from the neutron star was the surrounding shell of silicon, from which it created large amounts of ^{56}Ni. Nuclei with 56 nucleons, like nickel, cobalt and iron are amongst the most stable, so far as the strong force which binds nucleons together is concerned: they lie at the minimum in the graph of nuclear potential energy per nucleon versus atomic number. Table 21 shows a calculation of the principal amounts of material manufactured by SN 1987A (Woosley *et al.* 1987).

Radioactivity in the supernova

On longer timescales than are available in a supernova explosion ^{56}Ni is unstable by beta decay (the weak force). It decays radioactively through ^{56}Co to ^{56}Fe. The decay goes first by electron capture with a half life of 6 days:

$$^{56}\mathrm{Ni}_{28} + e^- \rightarrow {}^{56}\mathrm{Co}_{27} + \gamma + \nu.$$

The cobalt nucleus decays to iron, by electron capture (half life 77 days):

$$^{56}\mathrm{Co}_{27} + e^- \rightarrow {}^{56}\mathrm{Fe}_{26} + \gamma + \nu$$

and, less importantly, by positron emission:

$$^{56}\mathrm{Co}_{27} \rightarrow {}^{56}\mathrm{Fe}_{26} + \gamma + e^+ + \nu.$$

Thus nickel is transformed to iron, which is stable and lasts indefinitely. The radioactive decay takes place in a two-stage chain.

Radioactivity has the remarkable property that a given quantity of nuclei halves in a period of time called the half life of the radioactive decay. This in effect means that the number n of nuclei falls exponentially:

$$n = n_0 \exp(-t/\tau),$$

τ in this equation is called the mean life of the radioactive nuclei. In a two-stage chain like the one above, the pace is set by the slower stage in the chain, and the decay of cobalt is the slower. This means that half the nickel created by SN 1987A on 1987 February 23 had converted to iron by 77 days later, i.e. by May 11; half the remainder (i.e. $\frac{3}{4}$ of the whole) had converted by July 27; half the remainder (i.e. $\frac{7}{8}$ of the whole) had converted by October 12; and half the remainder ($\frac{15}{16}$ of the whole) by December 28.

The half life of ^{56}Co at 77.12 days translates to a mean life of $\tau = 111.26$ days.

In each of the stages of the radioactive chain from nickel to iron the resulting nucleus is created in an excited state. The excited nucleus is very similar to an excited atom – just as an electron in an atom, which has surplus energy, gets rid of it by radiating light or ultraviolet radiation, the nucleus has surplus energy which it disposes of by radiating gamma-rays. These are nothing more than photons of particularly high energy (Table 22).

Just as an electron in an excited atom has particular energy levels to which it can drop, and so radiates photons of a particular energy, so too the nucleus can only drop from its excited state to particular levels and is restricted in the gamma-ray energies which it can radiate. Just as a collection of heated and excited atoms radiates a spectrum of light which contains emission lines at particular wavelengths and energies, so too a collection of excited nuclei produced by nucleosynthesis and radioactive decay radiates a gamma-ray emission line spectrum at particular energies. Just as the visible light emission

Table 22. *Photons in the electromagnetic spectrum*

Frequency (cycles/s)	Waveband	Wavelength	Energy
10^{26}		10^{-12} μm	1 erg
10^{25}		10^{-11}	
10^{24}		10^{-10}	
10^{23}		10^{-9}	
10^{22}	gamma-rays	10^{-8}	
10^{21}		10^{-7}	
10^{20}		10^{-6}	1 MeV
10^{19}		10^{-5}	
10^{18}	X-rays	$10^{-4} = 1$ Å	
10^{17}		10^{-3}	1 keV
10^{16}	ultraviolet	10^{-2}	
10^{15}	light	10^{-1}	
10^{14}	infrared	1 μm	1 eV
10^{13}	submillimetre	10	
10^{12}	radiation	10^{2}	
10^{11}	microwaves	1 mm	
10^{10}		10	
10^{9}		10^{2}	
10^{8}	short	1 m	
10^{7}	radiowaves	10	
10^{6}		10^{2}	
10^{5}	radiowaves	1 km	
10^{4}		10	
10^{3}		10^{2}	
10^{2}		10^{3}	

line spectrum of atoms is a characteristic fingerprint which uniquely identifies the atoms, so too the gamma-ray spectrum from radioactive decay uniquely identifies the nuclei responsible.

The upshot is that if the spectrum of gamma-rays produced in the radioactive decay of ^{56}Ni (see Table 23) could be observed in SN 1987A, this would be proof positive that it contained radioactive nickel. Since radioactive nickel has a life which is measured in hundreds of days, it could not have been created in some distant and hypothetical process – it could only have been made in the supernova explosion itself.

Table 23. *The emission line spectrum of gamma rays from radioactive nickel and cobalt*

Spectral line energy (keV)	Line intensity (%)
$^{56}Ni_{28}$: half life 6.10 d; decay energy 2.11 MeV	
163	99
276	35
472	35
748	48
812	87
1560	14
$^{56}Co_{27}$: half life 77 d; decay energy 4.57 MeV	
847	100
1038	14
1238	64
1771	14
2035	7
2598	14
3254	7

Detection by Solar Max of gamma-rays from SN 1987A

Gamma-rays are very energetic and, like X-rays, they can penetrate material to a considerable extent. However, when they pass into the top of the Earth's atmosphere they penetrate only into the upper layers: air stops gamma-rays, and so cosmic gamma-rays do not reach the surface of the Earth. Only by launching gamma-ray detectors above the atmosphere can they directly be seen. Balloons can carry gamma-ray detectors above the majority of the atmosphere, although since they need air to float on clearly balloons cannot reach free space. Balloon flights last a matter of hours or days; the longest flights occur at the all important 'turnaround opportunities' – the times when the upper atmospheric winds over the launch sites are reversing direction and therefore do not quickly blow the balloons out of radio range. Rockets, on the other hand, can carry gamma-ray detectors above the atmosphere for a brief flight of just minutes before returning to Earth. Space satellites can carry gamma-ray detectors above the atmosphere and into the solar system – their life can be measured in years.

Fig. 54. The satellite Solar Max measured the 847 keV gamma-ray intensity from the supernova throughout 1987. For most of the year, even when the supernova appeared at the end of February (arrow), the intensity detected from the area of SN 1987A was consistent with zero counts per second. But from 1987 August, there was a small but significant increase in gamma-ray flux from the region of the supernova: the last eight points are all above the zero line, with a mean level of 0.040 ± 0.009 847 keV gamma-rays detected per second (1.0 ± 0.25 gamma-rays per second per square centimetre).

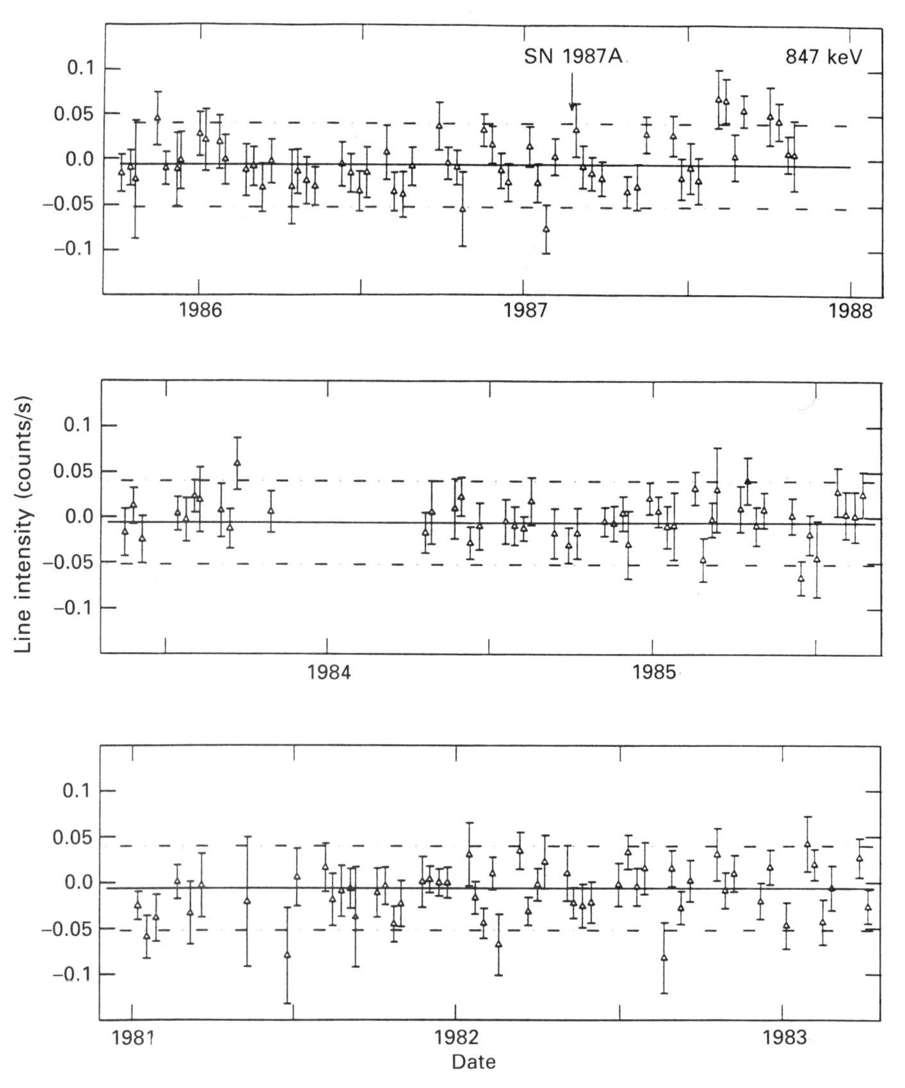

One such gamma-ray satellite is the Solar Maximum Mission – its initials are SMM but it is also familiarly known to its scientists as Solar Max. Scientists working with Solar Max realised in 1987 December (Matz *et al.* 1987) that they had detected gamma-rays from SN 1987A. As its name implies Solar Max was launched (in 1981 by NASA) to study the Sun, during its cycle of sunspot activity. With rare exceptions the satellite and its instruments are always pointed towards the Sun. The Sun moves around the ecliptic and the LMC happens to lie very near to the ecliptic pole. SN 1987A is therefore always within a few degrees of the direction at right angles to the Sun and to the forward viewing direction of Solar Max. The gamma-ray spectrometer contains seven gamma-ray detectors – 3 in diameter, 3 in thick cylinders of sodium iodide; the spectrometer is shielded by its sides against gamma-rays which do not come from the front. But nevertheless one third of the higher energy cosmic gamma-rays do pass through the shields, and constitute a background of cosmic events usually regarded by the solar physicists as a nuisance. Thus some gamma-rays from SN 1987A can be detected by Solar Max.

But, how can gamma-rays from SN 1987A be distinguished from gamma-rays from all the other cosmic sources if gamma-rays from any direction can penetrate the shields?

Solar Max orbits the Earth, and the LMC is periodically obstructed by the Earth as seen from the satellite. Not even the most energetic gamma-rays pass through the solid body of the Earth, and so the Earth acts as a shutter which periodically occults gamma-rays from SN 1987A. When the shutter is 'closed' the gamma-rays which Solar Max sees must be coming from the rest of the Universe; when the shutter is 'open' they may be coming from SN 1987A as well as the rest of the Universe. The difference represents the flux of gamma-rays coming from SN 1987A.

The Earth itself is a source of gamma-rays produced by interactions of cosmic rays with its atmosphere and a correction has to be made for the effect of the varying strengths of gamma-rays produced by the varying position of the Earth relative to the satellite as this shutter operates. It was possible to test how effective the correction procedures were by determining whether there were excess gamma-rays coming from positions separated in Right Ascension and Declination from SN 1987A – the Earth periodically occults stars at positions in the sky different from the supernova. No gamma-rays came from the test positions, only from the position of SN 1987A.

Thus, after all the tests and corrections were made, it became clear (Matz *et al.* 1988) that, since the beginning of 1987 August, Solar Max had observed gamma-rays from SN 1987A in the 847 keV and possibly also the 1238 keV spectral lines from the radioactive decay of ^{56}Co.

It had been proved that supernovae manufacture elements by nucleosynthesis.

Balloons detect gamma-rays from SN 1987A

In addition to the monitoring of gamma-rays from SN 1987A by Solar Max, several balloon experiments have been launched to detect them. Initial experiments were unsuccessful in detecting the supernova – but in fact it had been anticipated that they would not find gamma-rays in the early flights. As Guenter Riegler, director of NASA's supernova programme, put it: in the early flights, 'we were basically keeping the theoreticians honest' (Andersen 1988). It forced theorists to discuss the question about the first appearance of the gamma-rays and declare their theories before the detections.†

High altitude balloons launched on 1987 October 29 and November 18 from Alice Springs in Australia were the first to confirm the two cobalt spectral lines (Sandie *et al.* 1988, Cook *et al.* 1988), although bad weather in Australia in November curtailed several flights. Further flights successfully saw the supernova (e.g. Mahoney *et al.* (1988)). One flight carried a gamma-ray detector on the largest balloon ever launched in Antarctica (12 million cubic feet). The balloon was launched from McMurdo Station near Mt Erebus on 1988 January 8 and flew for 72 hours before the detector was released from the balloon by parachute, falling 360 km west of the Vostok research station. To be used again, the payload was recovered by a Lockheed LC-130 aircraft, which landed beside the downed detector. The aircraft rolled for 11 miles on its Teflon skis over the raw snow surface before gathering enough speed to take off again, with its cargo.

Balloon flights carry larger detectors than it is possible to launch in a satellite. But balloon flights do not last as long as satellite missions. They can give potentially more accurate measurements of some data at particular times. It is possible for instance that they might determine the width of the gamma-ray spectral lines. The width is caused by the Doppler effect due to the velocity of outflow of the supernova. When the velocity is known, it may be possible to determine where in the outflowing material the gamma-rays originate: if they

† The quotation is revealing. Identical words are used in the tactics of American football about the play by play decisions as the offensive side tries to prevent the defensive side from guessing its next move. Evidently the quote implies that Riegler was viewing science as an enjoyable contest between observers and theoreticians.

were on the outside of the explosion their speed would be larger than if they were on the inside.

Further evidence of the creation of nickel and the radioactive decay of cobalt came from X-ray observations from the Ginga and Kvant satellites.

Ginga

The third Japanese X-ray astronomy satellite, called Astro-C, was launched on a Japanese M-3SII rocket on 1987 February 5, three weeks before the appearance of SN 1987A. The satellite was renamed Ginga (Japanese for galaxy) after launch. On board was an X-ray detector known as the Large Area Counter (LAC), built by a consortium consisting of the Institute of Space and Aeronautical Studies, Tokyo, the Rutherford Appleton Laboratory and the University of Leicester. The satellite had the best sensitivity amongst the current detectors in the critical X-ray region. The satellite was made to scan the region of SN 1987A regularly from February 25, following an accelerated and curtailed test programme to bring it to operational readiness (see Chapter 2).

In 1987 September (Makino 1987a) the Ginga team suspected from observations made in August that a new X-ray source was present in the LMC. In October (Makino 1987b) they suggested that the new source was the supernova. In retrospect, the Ginga team (Dotani *et al.* 1987) had observed X-rays from the new X-ray source appearing from mid-June onwards. Taking account of the known X-ray sources in the LMC, they concluded that they had been watching the gradual appearance of the supernova. The X-rays were variable at first, increasing in intensity up to mid-January 1988. Then they began to fade.

Kvant

The Mir ('Peace') space station has been orbitting the Earth since 1986 February 20 in the longest space mission ever undertaken. The name Mir refers to the core of the space station; it has six docking ports, four on the sides and one at each end (the 'axial docking units'). On 1987 February 5 the manned spacecraft Soyuz TM-2 carried to Mir Cosmonaut Yuri Romanenko, a veteran on his third space-flight, and Cosmonaut Alexandr Laveikin. They docked Soyuz TM-2 at the 'front' axial unit, and occupied the Mir space station, re-activating it after it had been vacated in 1986 July.

The space station cannot only be vacated and re-occupied: it can also be re-configured by docking scientific modules at the unoccupied ports in response to changing scientific requirements. On 1987 March 31 the first of the scientific modules was sent to Mir. Its name was Kvant ('Quantum'). It carried gyroscopes for orienting the Mir space station, a new type of life support system for replenishing oxygen and new solar panels to power Mir; it also carried two astronomical observatories: one for ultraviolet astronomy, called Glazar, and one for X-ray astronomy called Röntgen. Röntgen consisted of four instruments, built cooperatively between the Soviet Union and western European nations:

> Sirene-2 is a Gas Scintillation Proportional Spectrometer provided by ESTEC on behalf of the members of ESA. This X-ray telescope is intended to gather spectra of X-ray sources in the 2–100 keV range.
>
> TTM is a Coded Mask Imaging Telescope, built by the University of

Fig. 55. The X-ray light curve of SN 1987A as observed from Ginga shows the first detection 130 days after outburst and a more or less constant intensity, except for a curious sharp 'flare' in 1988 January.

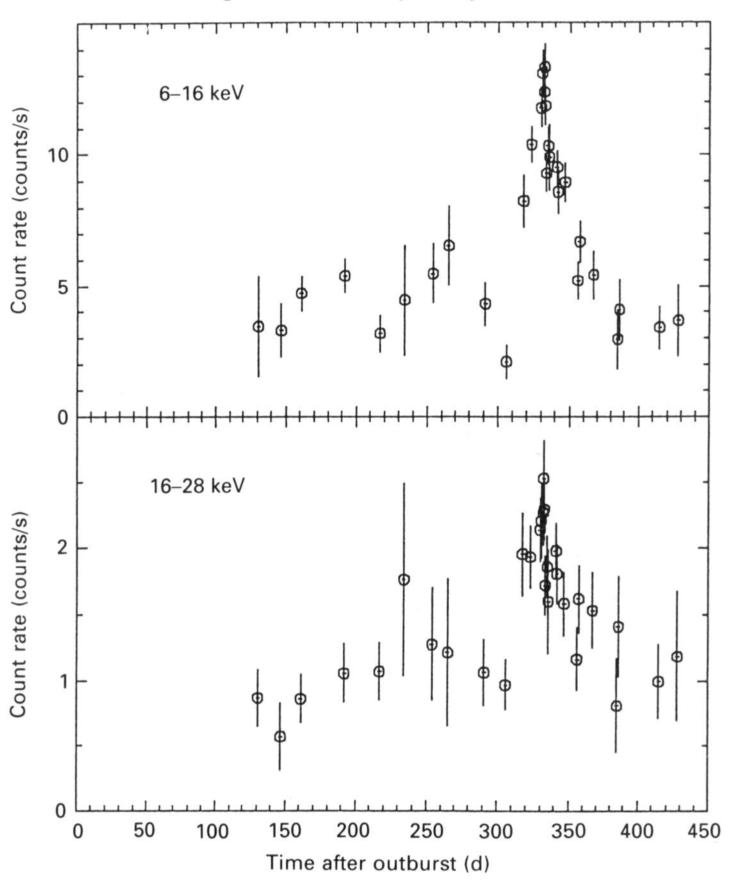

Birmingham and the Space Research Laboratory, Utrecht. This X-ray detector contains a rotating mask with holes which interrupt the beams of X-rays from sources in a 15° region of the sky. From the periodicities in the signals detected in the telescope, the positions of the sources can be reconstructed, and the telescope gives an X-ray image of the sky in the region. It detects X-rays from the 2–30 keV range.

HEXE is a phoswich scintillation spectrometer, built by West Germany's Max Planck Institute for Physics and Astrophysics, to detect high energy X-rays in the region 15–200 keV and study their variability.

Pulsar X-1, from the Soviet Institute of Space Research, is for studying X-ray and gamma-ray sources to very high energies (20–800 keV).

Kvant rendezvoused with Mir on April 14, and made two unsuccessful attempts to dock with Mir's rear axial docking unit. On the first occasion the control system of Kvant failed to find the docking port and Kvant sailed past Mir at a distance of only ten metres! On the second occasion, a correct soft docking

Fig. 56. The X-ray spectrum from SN 1987A, observed by Ginga, is quite unlike any other X-ray source (data from 1987 September 3). The fact that the spectrum has two components is shown by the abrupt kink in the spectrum at an X-ray energy of 12 keV. The soft component decreases in intensity from 4 to 12 keV. The hard component has a flat spectrum from 12 to 35 keV.

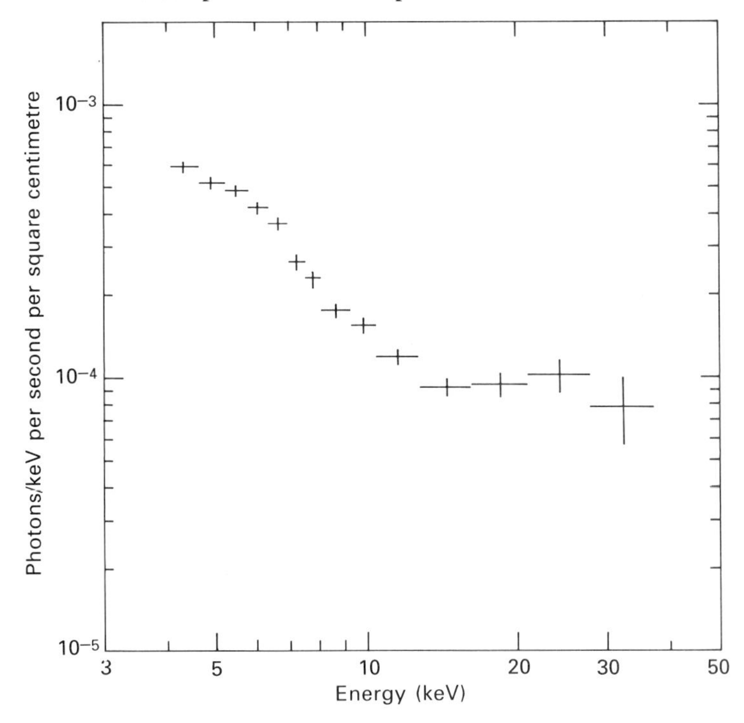

was made but the Kvant module would not lock on to Mir. Romanenko and Laveikin donned space suits and went outside on April 11 to rescue the Kvant module. Pulling Kvant apart from Mir, they found a small object in white wrapping between them, preventing the two from proper contact. This object was never identified, since after it was cut away it floated into space where it presumably still orbits the Earth amongst the many thousands of pieces of space junk; it could have been a protective covering or bag mistakenly left on Kvant by the ground crew.

Kvant was then successfully mated with Mir. On April 23 Kvant's propulsion unit was detached, and two days later the first of a series of further modules called Progress 29, carrying fuel for Mir, was attached where the propulsion unit had been. The modular space station, in which Soyuz TM-2, Mir, Kvant and Progress 29 had been assembled in a row, began its main programme. After a lengthy series of tests of the controlling gyroscopes, the cosmonauts could orient the space station in the direction required for the observatories to see the target; subsystems in the instruments carry out any necessary fine pointing.†

The first observation by the Röntgen observatory of the region of Supernova 1987A was made on 1987 June 8; no X-ray signal from the supernova was detected. The supernova was first detected on 1987 August 10 by two of the X-ray detectors: Pulsar X-1 and HEXE. It was first detected by the high-energy X-ray experiment (HEXE), and its spectrum was extended to higher energies by the

† The cosmonauts also maintain equipment in the modules. In the case of Kvant cosmonauts Titov and Manarov attempted, in an unscheduled spacewalk on 1988 June 30, to repair the TTM, replacing its detector. But a wrench snapped as it was being used to pry off the brass clamp holding the detector in the telescope, and the repair was unsuccessful.

Fig. 57. After the Soyuz TM spacecraft had arrived with two astronauts who re-activated the Mir space station, the Kvant module docked at the other end. Diagram courtesy of P. Willmore (Birmingham).

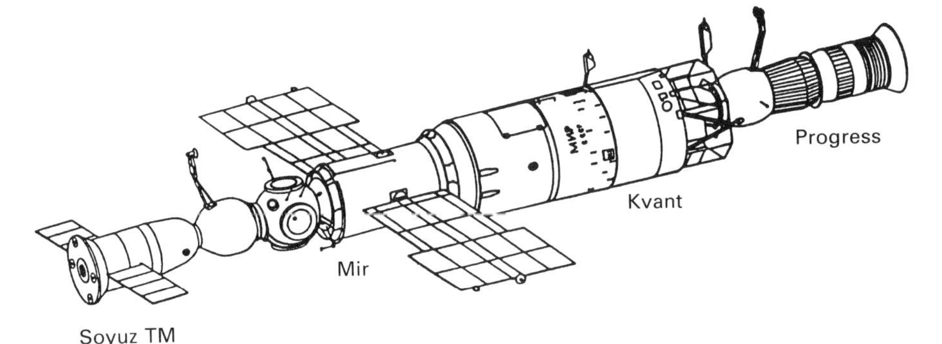

Pulsar X-1 detector. It was, however, never significantly detected by the TTM instrument before it broke. The supernova became the main target of the Röntgen observatory (Sunyaev *et al.* 1987, Skinner *et al.* 1988).

X-rays from the supernova

The X-ray spectrum of SN 1987A (Masai, Hayakawa, Itoh and Nomoto 1987, Itoh *et al.* 1987) consists of soft, thermal X-rays (energies less than 10 keV) which perhaps arise from material heated by the shock-heated ejecta rushing outwards from the supernova and colliding with circumstellar material, plus a more unusual hard component (above 10 keV).

The harder X-rays come from degraded gamma-rays produced by the radioactive nickel and cobalt. The gamma-rays first encounter electrons in the

Fig. 58. The spectrum of SN 1987A from the X-ray instruments on the Röntgen observatory (crosses, squares and diamonds) is consistent with two calculations (histograms) of the spectrum produced by gamma-rays which are degraded by interaction with the supernova envelope to form X-rays.

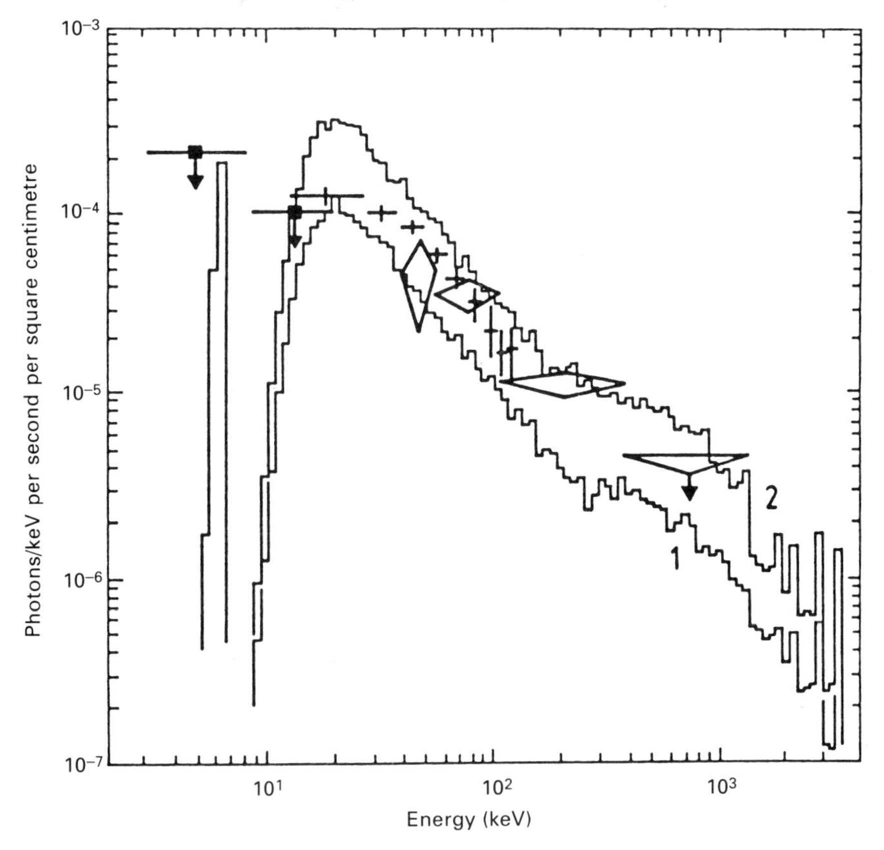

envelope material. They ricochet off the electrons in a process known as Compton scattering. Some of the gamma-ray energy is given to the electrons, heating them. The residual gamma-rays, which are less energetic than they were, may ricochet again and again; they gradually lose so much energy that they can better be called X-rays. The X-rays not only take part in the same degrading process but also are absorbed by the electrons in the ions of heavy atoms in the envelope material (photoelectric absorption). The X-rays emerge from the surface layers when the envelope becomes thin enough.

A natural explanation for the hard X-rays from SN 1987A seems thus that they originated in the gamma-rays produced by the radioactive cobalt. The hard X-rays from SN 1987A are thus also evidence for nucleosynthesis in the supernova.

Fig. 59. Contour map of a 2° square region of the LMC surrounding SN 1987A, produced by the TTM X-ray telescope on Kvant before it failed. Two known X-ray sources are marked LMC X-1 and PSR 0540-693 (the X-ray emission from these has to be accounted for in observations of SN 1987A). The other contour levels in the picture are consistent with noise in the telescope, including the patch which is just offset from the position of SN 1987A. University of Birmingham data.

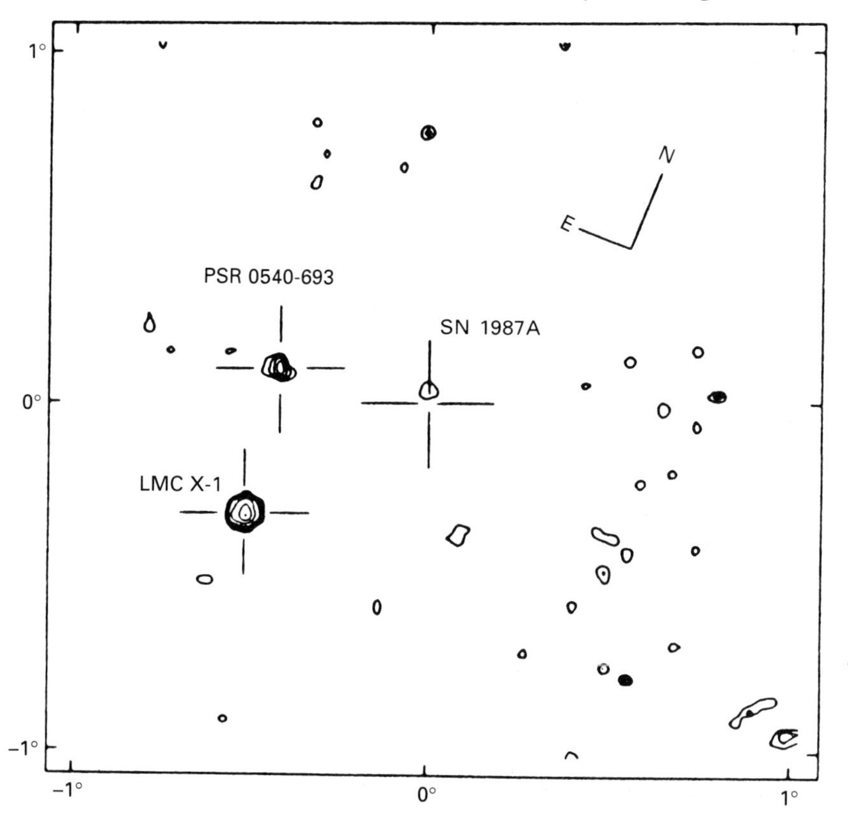

Radioactivity feeds the light curve of SN 1987A

Radioactive nickel was produced by the supernova in the mantle of material which surrounded the core. This mantle lay deep within the envelope. Only when the envelope became transparent to gamma-rays and X-rays could they escape from the envelope and cross space from the LMC to us, bringing direct news of the nucleosynthesis which had taken place in the supernova. This is why there was a delay in the discovery by Solar Max of the gamma-rays from cobalt: the gamma-rays were cloaked from us between 1987 February 23 and the beginning of August (Figure 54, p. 160). However, astronomers had even before then become convinced from indirect evidence that the supernova had created radioactive nickel. Indeed, they had measured how much of it had been created – 0.08 solar masses!

While the envelope was opaque to them, the gamma-rays produced by the radioactive nickel and cobalt interacted with the material in the envelope. The gamma-rays degenerated into heat which emerged from the supernova in several different forms. First to be recognised were infrared and light radiation.

The infrared and light photometry of SN 1987A, as described in Chapter 9, can be converted into a measurement of the total luminosity L ergs/cm^2 per second of the supernova. For reasons which it is not necessary to go into astronomers use a logarithmically related quantity called the bolometric magnitude, m_{bol}. It is related to L by:

$$m_{bol} = -2.5 \log(L) - 11.50.$$

When m_{bol} was plotted against time, astronomers found that from 1987 July until 1987 November it produced a straight line. This was easy to relate accurately to the heating of the envelope by the gamma radiation from the radioactivity. If all the heat entering at the bottom of the envelope material exits from the top as light and infrared radiation, the luminosity L follows the exponential decay of the nickel:

$$L = L_0 \exp(-t/\tau).$$

Putting this expression into the definition of bolometric magnitude we get

$$\begin{aligned} m_{bol} &= -2.5 \log[L_0 \exp(-t/\tau)] - 11.50 \\ &= -2.5 \log(e) \ln[\exp(-t/\tau)] - 2.5 \log(L_0) - 11.50 \\ &= 1.09t/\tau - 2.5 \log(L_0) - 11.50 \end{aligned}$$

This is the equation of a straight line, when m_{bol} is plotted against time; its slope is connected with the mean life τ of the radioactive decay and the intercept at $t=0$ is connected with L_0 which in turn is connected with the total mass of nickel originally created. From the slope of the line, Feast (1988) deduced a

Fig. 60(*a*). From day 140 after outburst until day 265 the magnitude of SN 1987A, including the whole spectral output fell in a straight line with time. From day 265 the output began to dip below the straight line. SAAO data.
(*b*) The accuracy of the straight line portion can be seen in the magnified blow-up in the lower panel.

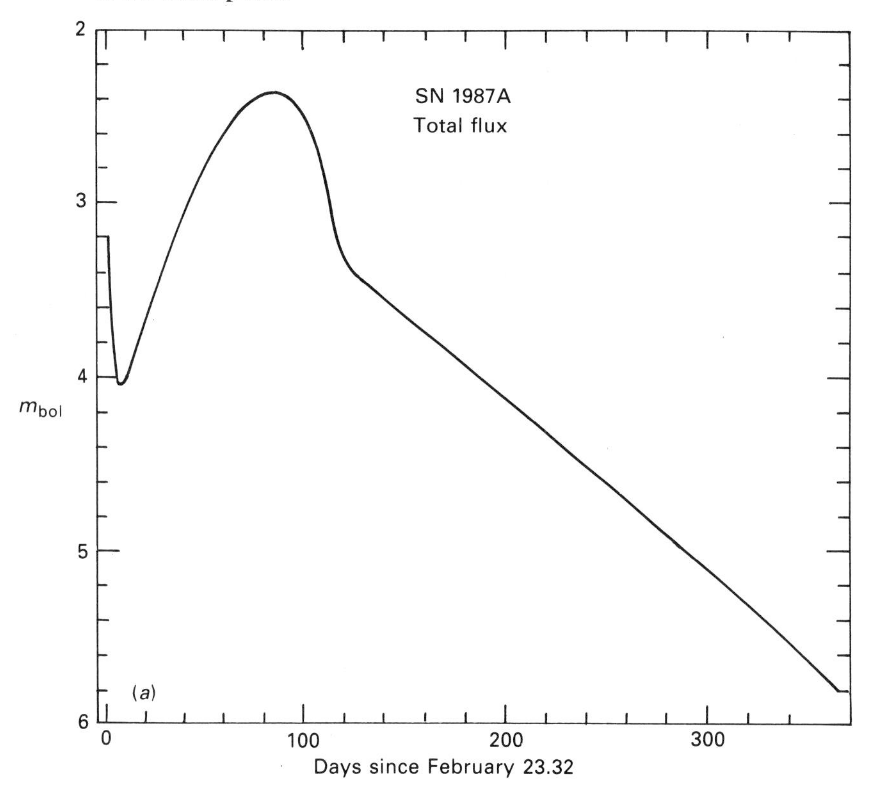

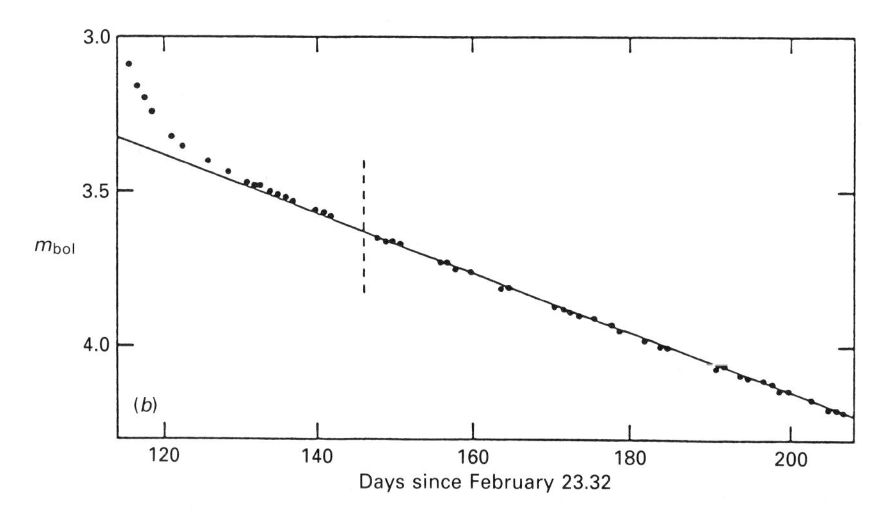

mean life of 110 ± 2 days. The value depended mainly on the assumptions which went into determining m_{bol} from the photometry but was in astonishingly good agreement with the mean life of cobalt, which is 111.26 days. The mass of nickel created was 0.077 of a solar mass ($\frac{1}{13}$ of a solar mass).

What happened after the end of 1987, when the decay of the infrared and optical curve fell faster than the expected exponential? Taking account of the gamma-rays and X-rays, which are also produced from the radioactive decay of cobalt, brought the observed energy decay back up to the exponential curve. The radioactive decay law has been followed well, at least up to 1988 August. In fact, it has been so well followed that no more than 5% of the supernova's power in 1988 August could have come from any source apart from radioactivity. There was no room in the observations for a second powerful source of energy. This had become an embarrassment to those theoreticians who had predicted that SN 1987A had produced a pulsar and who had expected that it would be powerful (Chapter 12).

The infrared spectrum of SN 1987A

The Kuiper Airborne Observatory (KAO) is housed in a Lockheed C141 military transport aircraft belonging to NASA. Up front sit the pilots in the cockpit. Towards the rear in the main body of the aircraft sit the Mission Director and the astronomers with their analysing equipment. Between is the 91 cm infrared telescope, viewing the sky from 41 000 ft. The aircraft flies well above the absorbing effects of the molecules in the Earth's atmosphere on cosmic infrared and submillimetre wavelength radiation. It flies above 99% of the Earth's water vapour, the principal absorber of infrared radiation in the submillimetre band (Plate 3).

In 1987 April, the KAO had been scheduled to fly a series of missions from Christchurch, New Zealand, in order to study Comet Wilson. It was easy to make a small change of target to a more distant southern hemisphere celestial object, far outside the solar system. The comet-watchers turned their attention to SN 1987A (Larson *et al.* 1987, 1988). Eight further missions to study the supernova were flown in 1987 November, covering the infrared spectrum from wavelengths of 1.5–100 microns.

Comparing the spectra between 4 and 12.5 microns, Rank *et al.* (1988) noted that a striking difference had developed in the supernova during the summer. In April the spectra showed the spectral lines of the Paschen and Brackett series; these are spectral lines from hydrogen, and are the infrared equivalents of the more familiar Balmer series in the optical region. Just as the supernova showed the Balmer emission which identified it as a Type II supernova, so it showed the infrared equivalents.

But by November, the supernova showed strong emission lines from heavy elements such as argon, nickel and cobalt. The mass corresponding to the observed spectral lines was 0.002 solar mass of nickel and approximately 0.008 solar mass of cobalt. The observation took place 250 days after the supernova outburst, and virtually all the radioactive nickel created in the supernova would have decayed to cobalt; the nickel observed was the residual stable forms of nickel produced amongst the by-products in the nucleosynthesis. But the cobalt probably was the radioactive cobalt produced by the decay of the radioactive nickel. After 250 days, with a mean life of 111 days, the amount of radioactive cobalt had declined to a fraction $e^{-250/111} = 1/10$ of the amount, 0.08 solar masses, created in the supernova. These calculations seemed to tally with the amount seen.

The airborne infrared results tie in with results from the ground. Iron produced from the cobalt decay appears in infra-red spectra obtained by ESO and by the AAT at the end of 1987. To make these spectra the effects of atmospheric absorption have to be cancelled out through careful observations of nearby stars whose spectra are relatively featureless, but are, of course, affected by the Earth's atmosphere. When a map of the atmospheric features has been obtained, their effects can be cancelled from the observations of the supernova. The results reveal the infrared spectrum of the supernova from 1.06 microns wavelength to 4.1 microns, with a couple of gaps where the atmospheric absorption is too much, and virtually all the infrared radiation from the supernova is absorbed in the Earth's atmosphere. The Anglo–Australian observing team (Allen 1988) observed a spectral line at 1.258 microns attributed to iron, probably produced by the radioactive decay of ^{56}Ni. The cobalt line died away while the iron line increased in strength, at just the right rate to match the radioactive decay (Meikle 1988).

Thus the core material had become directly visible in the infrared as well as in gamma-rays.

Radioactivity revealed – too soon?

Four independent kinds of observations showed that new radioactive elements had been produced in SN 1987A: observations of gamma-rays, which conclusively identified radioactive cobalt by its spectrum; observations of the light and infrared from the surface of the supernova, which pointed clearly to radioactive cobalt from the rate of decline of the light curve; observations of the infrared spectra of the supernova, which were consistent with an excess of radioactive cobalt and stable iron causing the emission lines; and observations of the X-rays, which were interpreted as a degraded gamma-ray spectrum of some radioactive element or other. All four observations showed that energy from the radioactivity became visible at the surface of the supernova in the summer of 1987.

The detections of the gamma-rays and X-rays were strikingly early. Theoretical calculations made before the detections indicated that the radioactive elements made near the core of the supernova would not become visible until late 1987 or early 1988. It was only at that time that the envelope of the

Fig. 61. David Allen of the AAO obtained this infrared spectrum with the AAT. He plotted the data from his night's observing in 1988 and hurriedly annotated the graph before displaying it on the AAT noticeboard the following morning for everyone to see. (Scientists usually like to make everything perfect, correcting slips of the pen, neatly lettering the labels and smoothing glitches in data, like an actress making up to present an immaculate face to the world. This 'working' graph is like an actress at rehearsal.) Allen notes that the peak at 1.26 microns signifies the presence of a strong spectral line from iron created by the radioactive decay of nickel and cobalt.

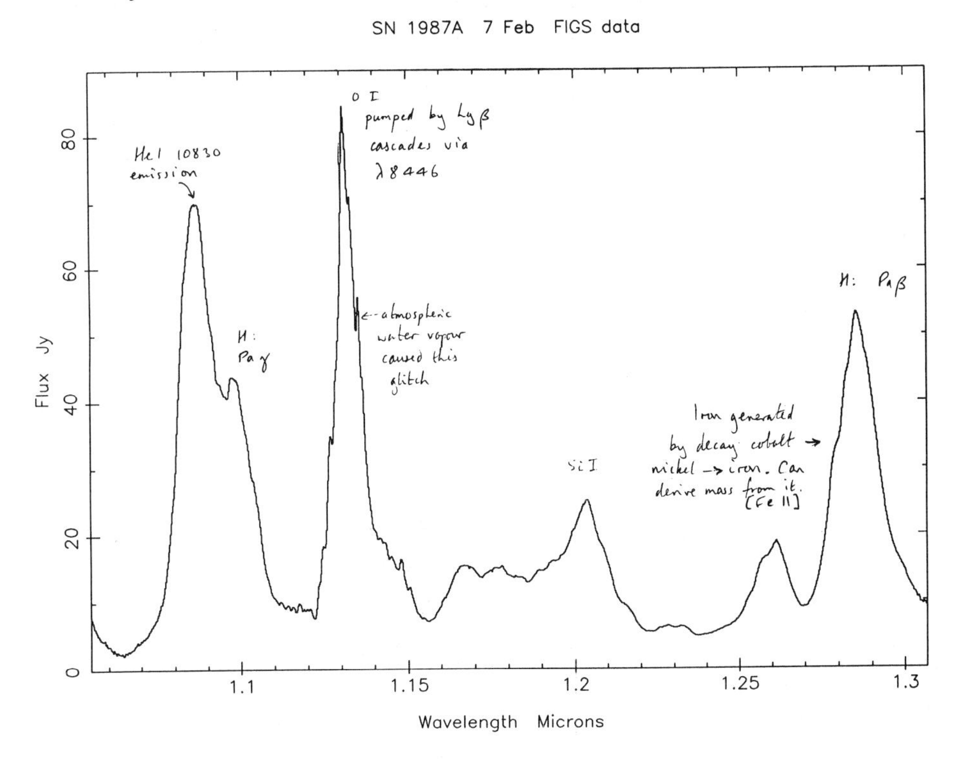

supernova would have thinned enough to reveal gamma or X-rays from the innermost parts of the supernova.

These calculations had been made on the assumption that the radioactive nickel was created in the centre of the supernova envelope, and stayed there, becoming cobalt, and that the envelope expanded uniformly. If the nickel and cobalt had been thrown towards the surface of the supernova envelope, by some mixing process, then the radioactivity would appear early.

Itoh *et al.* (1987) calculated what was necessary on the assumption that Sk – 69 202 had a mass of 11 solar masses when it exploded (about 9 solar masses was reckoned to have been lost in its stellar wind before the explosion). Supposing cobalt had been mixed to the outer edge of the oxygen-rich core material (2.4 solar masses of material lies in this core), then the X-rays observed by Ginga would not have become visible until the end of 1987. If the cobalt had

Fig. 62. Theoretical calculations of the intensity of the gamma-rays likely from SN 1987A showed how the flux of gamma-rays would quite quickly rise as the supernova's envelope became transparent to gamma-rays, and then decline expontentially (straight lines in the logarithmic plot) as the radioactivity decayed. Curves (*a*) and (*b*) are two calculations for an energetic and a less energetic explosion of a 15 solar mass star (the more energetic explosion makes more nickel and clears the envelope faster, so (*a*) lies above (*b*)). The data point shows the result of the first three months' detection of SN 1987A (average of the last eight points of Figure 54). The theory predicted a credible maximum level for the gamma-rays seen from Solar Max, but the gamma-rays appeared several months too early.

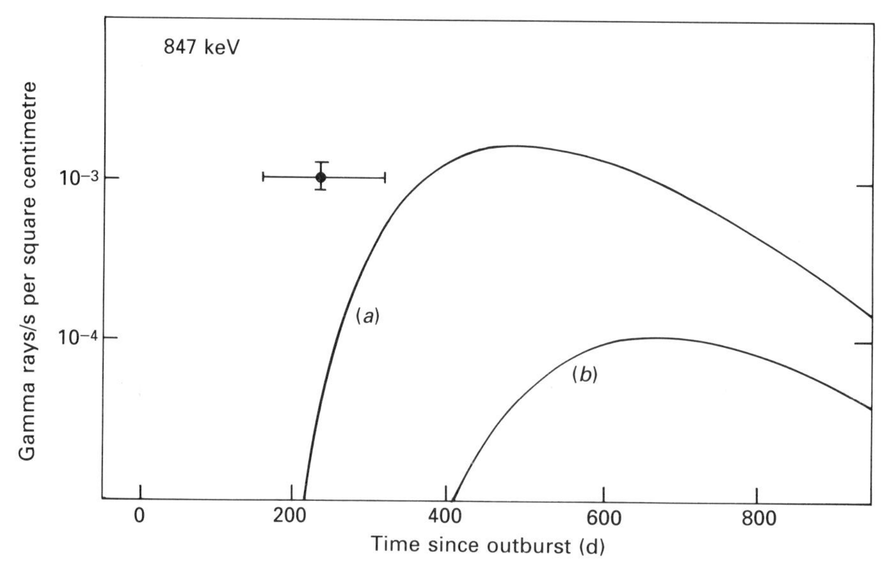

penetrated to the edge of the helium layer (4.6 solar masses lies inside this region), then X-rays would have appeared in about September. Only if the cobalt had mixed outside the inner 6 solar masses, and penetrated well into the hydrogen envelope, with over half the mass of Sk – 69 202 below this zone and less than half above, could the X-rays have become visible early in August.

What is being visualised by this 'mixing' process is that the nucleosynthesis process which created the nickel gave it kinetic energy so that 'fingers' of radioactive nickel squirted into the surrounding envelope like the pseudopodia of an amoeba. According to the first calculations of this kind of process by a Kyoto University group (Nagasawa, Nakamura and Miyama 1988), lumps break off the fingers in the first hour of the explosion and stir up the material behind the shock (Plate 7).

The material may be stirred up initially, but the outflow quickly smooths the turbulent material out into a uniform flow: like a marathon race, there may be chaos at the start, but the faster runners quickly take the lead, and the slower runners mostly never catch up.

Measurements of the broadening of the gamma-ray spectral line from cobalt decay are potentially capable of checking whether the positions for the cobalt determined by Itoh *et al.* (1987) are consistent with the velocity of the outflowing material. There are no consistent results on the Doppler shifts of the gamma-ray lines although some measurements claim widths corresponding to 1900 km/s (Mahoney *et al.* 1988) and 6000 km/s (Teagarden 1988).

Another possibility, not inconsistent with the first, was that maybe the supernova envelope had not expanded uniformly but had broken up in lumps. It would then have been possible to see past the lumps to part of the radioactivity within, just as blue sky may be glimpsed between clouds but not through a uniform fog. There is some evidence that this contributes to the true explanation of how the cobalt has become visible in the interior of the star.

Not only have the X-rays revealed themselves too early, but they did not fade away fast enough, compared with the predictions. By 1988 February 95% of the cobalt nuclei had decayed but the hard X-ray flux was the same as it had been since 1987 September. Only in Spring 1988 did the flux drop. Evidently the radiation flux increased as the supernova became more transparent and revealed more and more of deeper layers containing more cobalt, compensating for the decay of the amount of cobalt itself (Sagdeev 1988). Able to imagine any kind of distribution in the radioactive cobalt, theoreticians have been able to model the gamma-ray and X-ray 'light' curves quite successfully throughout 1987–8.

The future light curve

Of course ^{56}Ni and ^{56}Co were not the only radioactive elements produced by SN 1987A. ^{57}Co was also produced (half life 270 days) and ^{44}Ti (half life 48 years). After about 1990, the radioactive power from ^{57}Co will dominate the light curve for a few years, giving way to the ^{44}Ti; this will keep the optical and infrared brightness of SN 1987A at about magnitude 20 for a 100 years (McCray, lecture at the IAU General Assembly in Baltimore, 1988). It is going to be difficult to see faint details in SN 1987A – such as a faint optical pulsar – against the persistent radioactive glow.

The future of the elements created

What will happen to the new elements created by SN 1987A?

The cobalt will within a few years have all changed to iron. Eventually the iron and the other elements created in the supernova, and indeed in Sk – 69 202 beforehand, will be dispersed in the outrushing explosion and will contribute to the interstellar medium in the surrounding region of space in the LMC. Some of the iron will, as it collides with circumstellar and interstellar material in the LMC, be brought to a stop; mixed up with the general material, it will provide the raw material for the manufacture of, for example, terrestrial planets around some future, yet-to-be-formed star. In the same way iron in the centre of our own Earth was provided by some past and never-to-be-identified supernova early in the history of our Galaxy.

Some of the supernova ejecta will leave the LMC. The velocity of escape from the LMC is less than 100 km/s, and so any of the supernova material which does not slow down below this speed will enter intergalactic space, where it could pass from that galaxy into another – ours perhaps.

Supernovae, in general, help to progress the evolution of galaxies by enriching their interstellar material with the heavier elements. There are other enrichment processes such as the stellar winds which blow partly processed material from the surface layers of stars, and as these are lower velocity, the material always remains in the parent galaxy, even if it is a light galaxy like the LMC. So supernovae are not the only way that the interstellar material in a galaxy will become enriched; but light galaxies will, in general, not become as enriched so quickly as more massive galaxies, because the ejecta from supernovae can leave a light parent galaxy. This seems to be the reason why the

LMC manifests less metal content than our own Galaxy – its content of metals is just one quarter that in our own Galaxy.

The metallicity of the LMC seems to be connected with the reason that SN 1987A was unusual in occurring in a blue supergiant and in being so faint.

Why was the supernova comparatively faint?

Even though the supernova reached second magnitude and thus equalled the total brightness of the LMC, its parent galaxy, it is regarded by astronomers as fainter than usual. Astronomers took some time to realise that the very fact that Sk−69 202 was a blue supergiant explained why the supernova had not got as bright as supernovae usually do. They should have realised immediately, but they had forgotten the fundamentals of the theory.

Astronomers compare the brightnesses of stars by their absolute magnitudes – the brightness which they would have at a distance of 10 pc (just over 30 light-years). Leaving aside the interesting question of what more dramatic events would happen if there actually was a supernova at a distance of 10 pc, a typical supernova at that distance would appear at magnitude −18 or −19 (brighter than the full Moon but fainter than the Sun). Calculations showed that, at magnitude 2.8 in the LMC, SN 1987A would only be magnitude −16 at 10 pc. The discrepancy is even more marked if, instead of considering the visual magnitude, we consider the blue magnitude of SN 1987A and other supernovae. Its magnitude at 10 pc would be −15 compared to −17.5 or −18.5 for the majority of Type II supernovae. Perhaps two or three magnitudes fainter does not sound very much but it means that SN 1987A was at least 10 times, and perhaps as much as a factor of 40 times, fainter than average (Filippenko 1988). Was there an explanation for this?

What happens when a supernova explosion takes place in a relatively small star is that a smaller fraction of the kinetic energy of the explosion gets picked up by the envelope of the exploding star and re-radiated as heat and light. Blue supergiants are about the same luminosity as red supergiants, but much hotter and so much smaller than red supergiants (although still large compared to our Sun and fully justifying the name 'supergiant'). It was perfectly natural that if the supernova had occurred in a blue supergiant, like Sk−69 202, smaller than a red supergiant, the supernova would be fainter than normal.

So why did the supernova explosion take place in the blue supergiant and not, as apparently is more usually the case, in a red supergiant?

A supernova explosion is triggered by the collapse of the core of a star when its nuclear energy runs out, whereas the surface appearance of the star (temperature and size) is a consequence of the properties of the envelope; and there is not much direct connection between the properties of the envelope and the state of the core. In particular, the size of a star depends on the traces of elements heavier than hydrogen in the envelope (Renzini 1987), and in the LMC there is less of these trace impurities than usual (a quarter of the amounts in the Sun). Supernovae in the Magellanic Cloud type of galaxy are rather rare, and we therefore have not recognised faint supernovae like SN 1987A before.

The faintness of the early part of the supernova light curve meant that the shape of the light curve was surprising for the first few months. But after the light curve settled into a regular decline of 0.01 mag per day, astronomers recognised the cause as heat from the radioactivity, passing out of the supernova envelope. And at that point the light curve became recognisable as fitting the majority of other Type II supernovae (Feast 1988).

So in simplification: the first four months of the light curve is powered by the shock wave passing through the envelope of the progenitor star and this depends on the size of the star, and so on the circumstances in which the star finds itself, varying from supernova to supernova. But all Type II supernovae make radioactive nickel and power the later straight line part of the light curve in the same way.

11
The neighbourhood of a supernova

Mass loss from stars

In Chapter 5, I discussed how stars whose mass was greater than about 7 solar masses end up as supernovae. What happens to the stars which are smaller than this is that they end up being white dwarf stars. These are evolved, hot, small, dense stars whose masses can be as large as 1.5 solar masses or so. Now, it is clear that a star of say 5 solar masses cannot become a white dwarf of 1.5 solar masses without undergoing a considerable change in mass; more than half the star must be lost before the transition is possible. Mass loss, in fact, is a common and dominant factor in the evolution of stars of all sorts.

The loss of mass from stars shows in some of the pictures of the many and beautiful nebulae which surround stars. Sometimes such nebulae surround stars by chance or by the loosest association; but sometimes the nebulae are produced by the loss of mass from the parent star itself. Planetary nebulae, for instance, are nebulae which surround stars which are akin to white dwarfs, and are on their way to becoming white dwarfs. They are called planetary because many of them look spherically symmetric, like planets, but some are toroids, like ring doughnuts, which surround the central stars. The nebulae represent the mass lost from the star in its evolution, lit up by the energy radiated from the central star.

As the winds blow from the surfaces of stars, the material of which they are composed cools. If the material is hydrogen from the surface layers of the stars it is ionised as it begins its journey. In this state it can be detected by radio telescopes; they detect true radio 'stars'. But more and more of the hydrogen atoms recombine as the material streams into the colder regions of space around. If the star later alters to become a hotter, brighter more energetic radiator it may re-ionise the material which then becomes visible again.

If the material includes carbon and silicon, then not only do the carbon and silicon ions recombine into atoms, but the atoms stick together to make molecules. Depending on the composition of the material, the lost mass might form molecules like C_2, CO, CH_3, SiO_2 etc. The carbon-based molecules may last for a considerable time in incomplete states, with unsatisfied chemical bonds, and larger and larger molecules may build up – indeed, they may form crystals and dust grains. Carbon-rich stellar winds become sooty, with graphite flakes and amorphous lumps; silicon-rich winds form sand.

The grains which form in this way, whether of carbon or silicates, trap some of the radiation from the star and become warm – or, at least, warmer than interstellar space – at temperatures in the region of 100 K. They radiate infrared radiation.

In order for mass to be lost from a star, there must be some pushing force which blows the mass off, in spite of the gravitational force at the star's surface which tends to hold the mass back. The pushing force is represented by the winds and storms which take place on the surfaces of stars. Even in our own Sun, which is by no means a violent star in this respect, giant prominences occasionally arch above the Sun's surface at times of solar activity. A steadier wind blows constantly from the Sun's surface producing cosmic rays which, spiralling onto the Earth's magnetic poles, produce aurorae etc.

By and large, the faster, potentially more powerful winds blow from the surfaces of the hotter stars. Thus, blue (hotter) stars have stellar winds with velocities measured in thousands of kilometres per second and red (cooler) stars have winds measured in tens of kilometres per second. Because the winds from blue stars move more quickly than the winds from red stars they fill a larger volume in a given time and are more rarefied.

On the other hand, stellar winds have their greatest effect in pushing mass off the star when the force of gravity at its surface is at its weakest – i.e. when the star is biggest. This tends to be when the star is cooler.

Thus, there are two classes of stellar winds: fast moving rarefied winds, by which mass is pushed off hot large stars; and slow moving denser winds, which push mass off cool large stars. Large stars are, in general, the bright ones – supergiants, like Sk – 69 202. Thus it is pretty safe to postulate that, before it exploded, Sk – 69 202 had undergone mass loss via stellar winds.

Whatever the speed of the wind, the material which it blows from a star drifts further and further from the star, filling a larger and larger volume. The density of the wind thus decreases outwards. If the wind speed remains constant, a constant flux of material in the wind crosses each spherical surface centred on the star. This means that the density of the stellar wind material falls as the square of the distance from the star – at twice the distance, the wind density has fallen to a quarter. This is an idealisation, but conveys the right idea of a progressively more and more diffuse cloud of material surrounding stars like Sk – 69 202.

Mass loss from Sk – 69 202

There was little direct evidence until the supernova explosion that Sk – 69 202 had lost any mass, although it seems likely from general arguments (Maeder 1987) that stars like it in the LMC lose considerable amounts in the course of their evolution – even as much as a half their original mass.

The lost material might have been hydrogen – if not very much material was lost and a thin layer only was stripped from the outer surface.

Or it might have been hydrogen and helium – if the mass loss included virtually all of the hydrogen layer and penetrated into the helium shell.

Or it might have included carbon or silicon – if the mass loss were really severe, and exposed the products of the helium burning.

The mass loss itself need not result in *all* of an outer layer being stripped away to reveal the abnormal composition below. Stars have 'cloudy' surfaces with convective cycles penetrating more or less deeply into the star, and it is possible for the convection to dredge up processed material from inside the star and bring it to the surface.

If the mass loss included significant amounts of processed material with abundant amounts of carbon or silicon then Sk – 69 202 could have been surrounded by a cocoon of solid grains – graphite, soot or silicates. As outlined above, dust grains like this would have been warmed by the stellar radiation and would have re-emitted infrared radiation. There is no indication in archival observations that Sk – 69 202 emitted any unusual infrared radiation. For example, the IRAS satellite did not detect an infrared source at the position of Sk – 69 202.

However, there is some indication that a considerable amount of the outer hydrogen envelope material had been lost: this evidence is in the form of archival spectra of the surface of Sk – 69 202 taken in the decade before it exploded.

The only spectra of Sk – 69 202 which exist prior to the supernova outburst are from objective prisms. Objective prism photographs of the sky are taken with a giant camera through a wedge-shaped piece of glass, which spreads the images of all the stars into little spectra. The details of the spectra are hard to discern because the spectra are so small, but large numbers of spectra are recorded at once so this is a good way to carry out a survey to look rather crudely at many stars in one shot. The most detailed spectra of Sk − 69 202 are from two instruments at the ESO, one plate in 1977 with the ESO Schmidt telescope (Gonzalez *et al.* 1987) and some others with an astrographic telescope in 1972–3 (work by L. and M.L. Prévot, reported by Walborn (1988)).

Both sets of spectra are a bit noisy and the picture is not entirely clear. They may indicate that Sk − 69 202 had stronger helium spectral lines than other comparable stars in the LMC. For instance, in the ESO Schmidt plate a spectral absorption line near 4026 Å which is due to both nitrogen and helium is strikingly stronger in Sk − 69 202 than in Sk − 67 78 nearby. From the more detailed but more limited astrograph data it seems that the helium line is the one which is primarily responsible for this difference. The natural interpretation is that the surface of Sk − 69 202 was richer in helium than normal, as if we were then seeing into the region in which hydrogen had been partially converted to helium, or at least that enough of the hydrogen had been lost for convection to reach into the helium layer and bring excess helium to the surface.

The objective prism spectra of Sk − 69 202 thus led to the same conclusion as the construction of synthetic spectra of SN 1987A by Lucy (1987); his conclusion was that SN 1987A was somewhat overabundant in helium (see Chapter 9).

There is some confirmation therefore for the thought that some or more of the outer hydrogen layer of Sk − 69 202 had been lost before the star exploded.

The circumstellar environment of Sk − 69 202

In Chapter 5, I presented evidence that after its life as a dwarf Sk − 69 202 had been first a blue giant above the Main Sequence and then a blue supergiant, next a red supergiant, and afterwards a blue supergiant again, whereupon it exploded. Thus Sk − 69 202 probably created several zones of wind around itself (Chevalier and Fransson 1987; Chevalier 1987).

In the outermost zone, there was a very fast, very rarefied wind from the first blue giant and supergiant phase. This wind formed a large bubble around the star, probably over 100 light-years in radius. Its effect was to clear away the interstellar material; what was an ambient density of about 1 atom per cubic centimetre of interstellar material became 1 atom per litre of stellar wind material instead.

In the next zone inwards, there was a slower moving denser wind from the red supergiant phase. The star was a red supergiant star for about 1 million years. The outer boundary of the 'red' wind thus lies tens of light-years from Sk − 69 202 with its density gradually increasing inwards to a denser inner boundary. The inner boundary of the red wind marks the place in the wind where the red supergiant phase of Sk − 69 202 stopped. This was something between 1000 and 100 000 years ago; the inner boundary of the red wind thus lies about 1 light-year from the star.

In the innermost zone, there occurred another fast moving, rarefied wind from the second blue supergiant phase. This 'blue' wind filled the inner cavity in the red wind. The blue wind rapidly overtook the slower moving red wind. The blue wind pushed the red wind from behind, compacting its dense innermost boundary making it even denser. The shock of the collision between the slow red wind and the fast blue wind trying to overtake it formed a shell whose density is in the region of 100 atoms per cubic centimetre. Probably the shell is clumpy, as the collision fragments the winds. It might even have condensed in places to form dust grains.

The circumstellar structure of Sk − 69 202 is summarised in Table 24.

Fig. 63. A schematic picture (which is more like a series of notes than a scale drawing) shows SN 1987A interacting with its own stellar wind. Diagram by P. Lundqvist and C. Fransson.

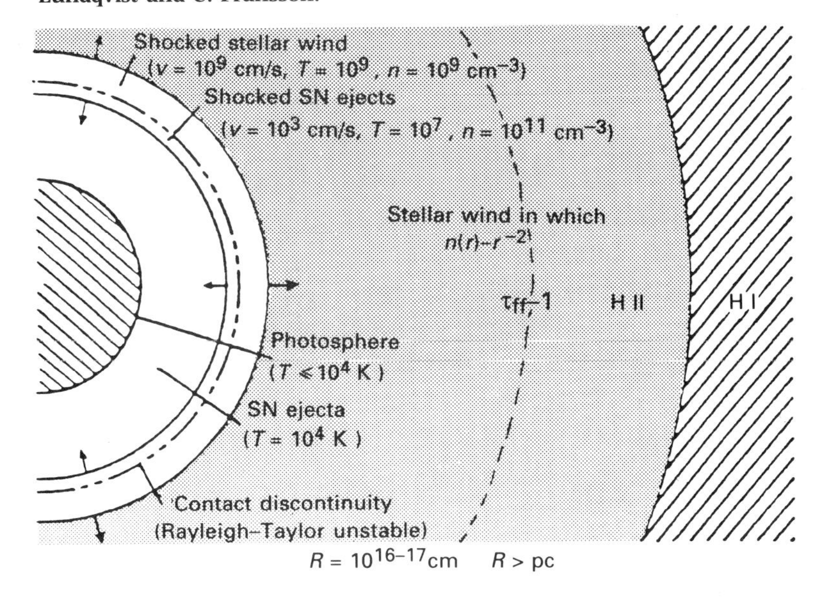

Table 24. *Circumstellar structure of Sk−69 202*

Zone	Position	Radial extent
blue wind	from star's surface to red wind	100 light-seconds–1 light-year
shell	at collision between blue and red winds	At 1 light-year
red wind	from shell to outer wind	1–50 light-years
outer wind	from red wind outwards	50–200 light-years
interstellar space	everywhere else	>200 light-years

The circumstellar shell appears

The first direct evidence for the existence of the circumstellar shell around Sk−69 202 was discovered in the supernova's spectrum in 1987 June, although, with hindsight, evidence was present in spectra taken on May 24 (Wamsteker *et al.* 1987; Kirshner *et al.* 1987; Cassatella 1987; Kirshner 1988b). This evidence took the form of narrow emission lines which appeared in the ultraviolet spectra of the supernova, obtained with the IUE.

Spectral lines of nitrogen, carbon, oxygen and helium were detected. They gradually increased in strength. The fact that the lines were narrow meant that they were not from fast moving material: otherwise they would have had a large width due to the Doppler effect of the velocity of the material projected onto the line of sight. The Doppler shift indicated that the material had speeds of less than 30 km/s, rather than speeds in excess of 10 000 km/s which would be typical of the outflowing supernova ejecta. The spectral lines therefore came from material in the supernova which had not (yet) been disturbed by the supernova explosion. This put the material beyond a distance of, say, 30 000 km/s × 3 months, i.e. 10 light-days from Sk−69 202.

In addition, the particular spectral lines which appeared were characteristic of relatively low density material.

Spectral lines are photons which are produced by electrons as they make jumps downwards between energy levels in the ions. Some spectral lines are called forbidden spectral lines because it takes a long time for the electrons to accomplish the transitions. In dense material an electron in the upper energy level may never get a chance to complete a forbidden transition because, even if poised to make the jump, it is interrupted when the ion is disturbed by an encounter with another one. In more rarefied material there may be longer intervals between encounters between ions and some electrons can make the so-called forbidden transitions. The forbidden spectral lines then do appear in the emission line spectrum.†

Forbidden lines had indeed appeared in the ultraviolet spectrum of SN 1987A, and so the density of the material from which they arose was relatively low. Two lines were the most crucial in estimating the density of the material. They were the semi-forbidden spectral lines of three-times ionised nitrogen (the symbol for this pair of spectral lines is NIV] at1483–6 Å). They are produced by material whose density is less than 10^5 electrons per cubic centimetre – much less than the density of the surface material of a star and more akin to the densities in circum- or inter-stellar space.

The idea was therefore that the ultraviolet light from the supernova (from the heating of the star's surface after the shock-breakout, perhaps) had ionised the circumstellar shell surrounding Sk – 69 202. It took some time – days or weeks – for the material to begin to recombine.

In addition it took extra time for spectral emissions from regions off to the side of the direct path of light from the supernova in our direction to make their extra journey towards us. As I write, somewhat over a year after the supernova occurred, there has as yet not been enough time for the parts of the circumstellar shell which lie a light-year behind the supernova to have responded to the ultraviolet flash and to have sent us the message that they have begun to recombine, since that takes at least two extra years of light travel time. Only regions of the shell lying nearest to us have had time to send us ultraviolet radiations informing us to the fact that they had been ionised by the supernova. Thus, as time has gone by more and more of the ionised shell has come within our horizon, and the strength of the spectral lines has increased. From estimates of the rate at which the strength of the emission lines has increased, it is possible to deduce that the shell is greater than 1.5 light-years in size. It contains about $\frac{1}{20}$ of a solar mass of material.

The nitrogen lines in the ultraviolet spectrum, although weak, were strong in comparison to the carbon and oxygen lines. In fact, given the weakness of the carbon and oxygen lines, the nitrogen lines ought not to have been detected at all. Nitrogen is calculated to be between 6 and 60 times more abundant than

† The symbol for forbidden spectral lines is to put the name of the ion which causes them in square brackets. The ion is represented by the name of the chemical element plus a roman numeral which is I if the atoms are neutral, II if it is once ionised, III if it is twice ionised, etc. Thus [NV] represents a forbidden line from nitrogen atoms which have lost four electrons. Sometimes a spectral line is produced by a transition which, although not long lived, is not rapid either: it is called semi-forbidden and represented by an incomplete square bracket.

normal when compared to carbon and oxygen – in the Sun, for example. This rules out the possibility that the narrow ultraviolet lines were formed in the general interstellar medium in the LMC, in which the nitrogen abundance is much more typically normal.

However, an excess of nitrogen is the result of the CNO cycle (see Table 5) of hydrogen burning in stars like Sk – 69 202. This would be consistent with the idea that it was the red wind, which produced the circumstellar shell. The wind had blown off dredged up material from the interior of Sk – 69 202, reaching into the helium shell but not through that into the carbon shell of the progenitor star.

All these details tended to confirm the identification of the IUE nebula as the circumstellar wind of the red supergiant from which Sk – 69 202 evolved.

Speckle interferometry

It would be very nice to have direct evidence for the existence of the circumstellar shell around the supernova – indeed, it would be a great advantage to have any pictorial evidence at all about the appearance of the neighbourhood of the supernova. This is difficult since, at the distance of the LMC, 1 light-year subtends an angle of 1″. Thus the circumstellar shell at the boundary of the red and blue winds lies only 1″ from the star. Because the Earth's atmosphere blurs star images to about 1″ in size it is impossible to form an image of the shell by conventional means – the shell would not be distinguishable from Sk – 69 202 on an ordinary photograph for example.

The technique of speckle interferometry overcomes, to some extent, limitations in the clarity of the Earth's atmosphere through which we have to view the Universe. The technique has given several puzzles for supernova astronomers to solve – and that is not surprising since this is the first supernova close enough to apply the technique to a case which could give an interesting answer. There are two versions of speckle interferometry which have been applied to SN 1987A – optical speckle interferometry and infrared speckle interferometry.

Optical speckle interferometry uses video pictures of a star recorded through a large telescope. Because of atmospheric scintillation, the star image in each video frame appears as a group of 'speckles', much like the fragmented image of a light bulb viewed through a pane of faceted bathroom glass. Each of the speckles represents the image of the star as seen through the same telescope without the presence of the atmosphere. Each speckle image within a frame can be correlated by computer, and the enormous number of frames recorded by the video camera can be summed in the computer. The end result is an image of the star with much better definition than by simply looking through the telescope.

Infrared speckle interferometry is similar, except that there is as yet no suitable imaging camera to record infrared radiation beyond about 1 micron wavelength. Instead, the positions of the speckles are recorded by scanning a slit rapidly across the image of the star and recording the infrared radiation with a photometer.

Fig. 64(*a*). If there is no atmosphere, the beam of light from a star can be focused by a telescope to a point-like image whose size is, in fact, set by the diameter of the telescope – the larger the telescope the smaller the size of the image.
(*b*) But if the telescope looks through a wavy atmosphere of air, the telescope forms a blurred picture of a star. It is no longer just point-like but is composed of numerous point-like 'speckles' in a large pattern. The speckles move quickly about the pattern as the wind blows the atmospheric waves over the telescope, and in a time exposure blur the image into the so-called 'seeing disc'. After a diagram by Peter Meikle (Imperial College).

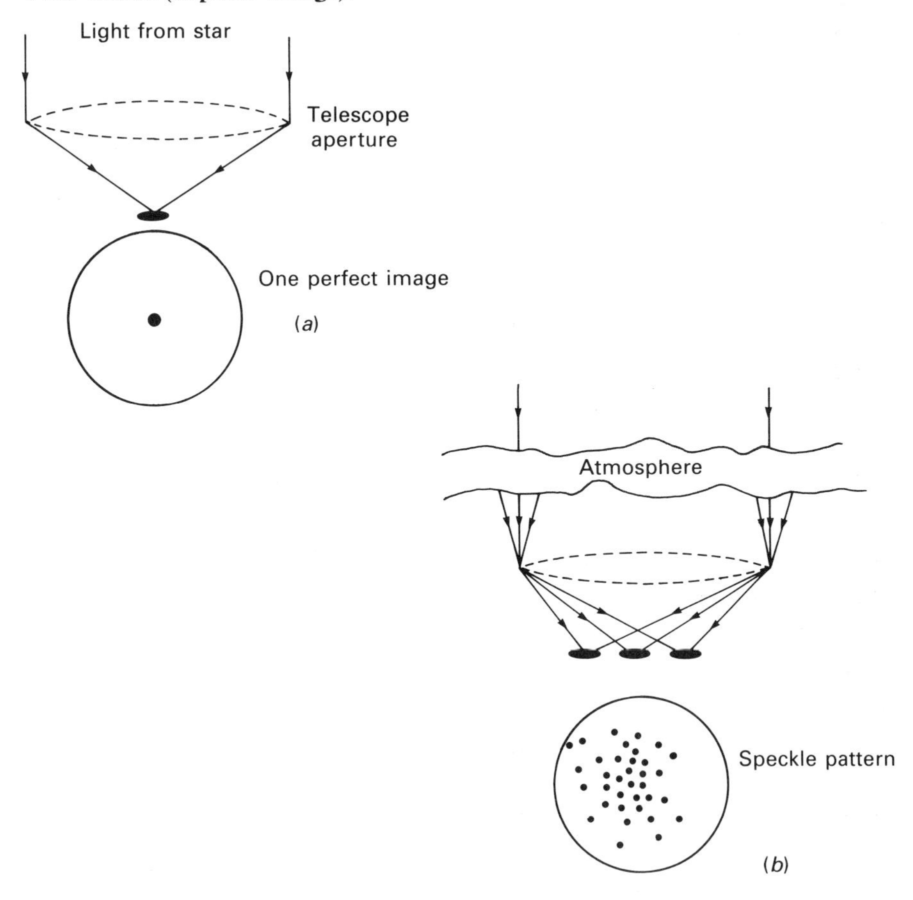

Infrared speckles

Infrared speckle interferometry (wavelengths of 2–5 microns) was applied to the supernova with the ESO 3.6 m telescope (Perrier *et al.* 1987; Chalabaev, Perrier and Mariotti 1988). Up to 1987 June 8 the supernova was indistinguishable from a point source, but from that time the supernova has appeared to have a surrounding halo. It is not easy to reconstruct from the scans of the speckles exactly what the infrared image of the halo looks like but it seems to consist of a hollow ring of radius 0.25 light-years. The most natural explanation seems to be that the infrared radiation was being re-radiated by

Fig. 65. If the image of a binary star is frozen on video tape, photon by photon (*a*), then each individual speckle in the pattern (*b*) can be discerned as the image of the binary star. Histograms of the separations of all possible pairs of photons in numerous video frames show an excess of pairs at displacements corresponding to the separation of the binary star (two of the possible histograms are represented in (*c*)). More intensive computer processing can reconstruct the actual image structure. After a diagram by Peter Meikle (Imperial College).

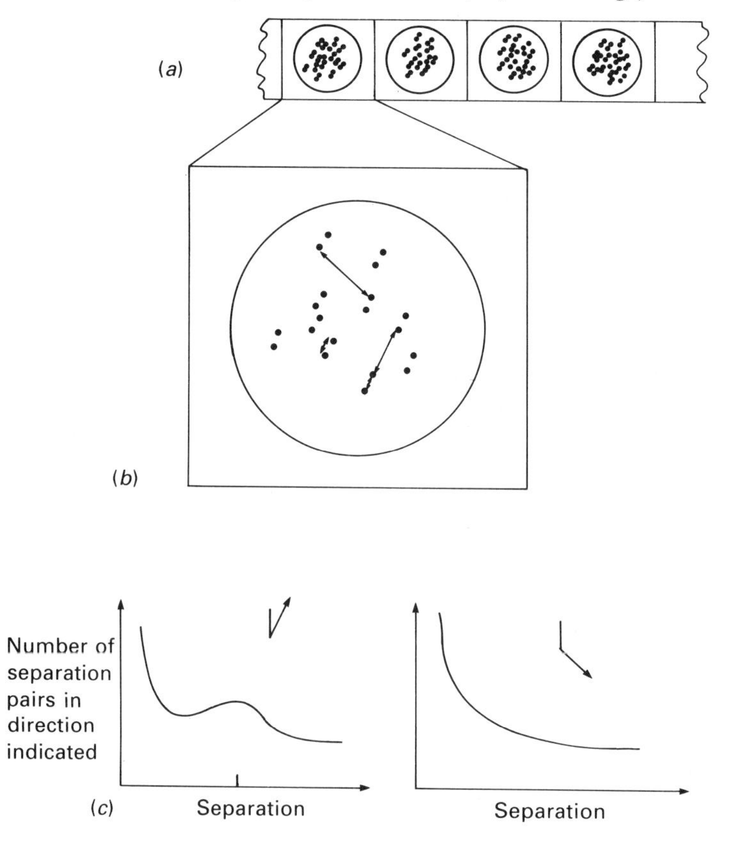

dust grains which had formed in the circumstellar shell between the blue and the red winds of Sk – 69 202.

This ring may be connected with the evidence in infrared spectroscopy for the presence of silicate grains. The infrared spectrum between 8 and 15 microns wavelength shows excess emission which represents warm (400 K) grains and spectral features ascribable to silicates (Aitken 1988). These grains cannot be packed in close to the supernova, or else they would completely obscure it. They fit naturally at the distance of the red supergiant wind.

Fig. 66. Speckle data for SN 1987A from the Imperial College speckle interferometer is shown as histograms of speckle separations at different angles ((*a*)–(*j*)); each member of each pair of histograms straddling a letter comes from different data samples. In the area of the diagram sidelined by black bars there is a hump on four adjacent histograms at a separation of 0.06″. Nowhere else in the diagram do four humps correlate like this and the observations confirmed the Center for Astrophysics' discovery of the Mystery Spot.

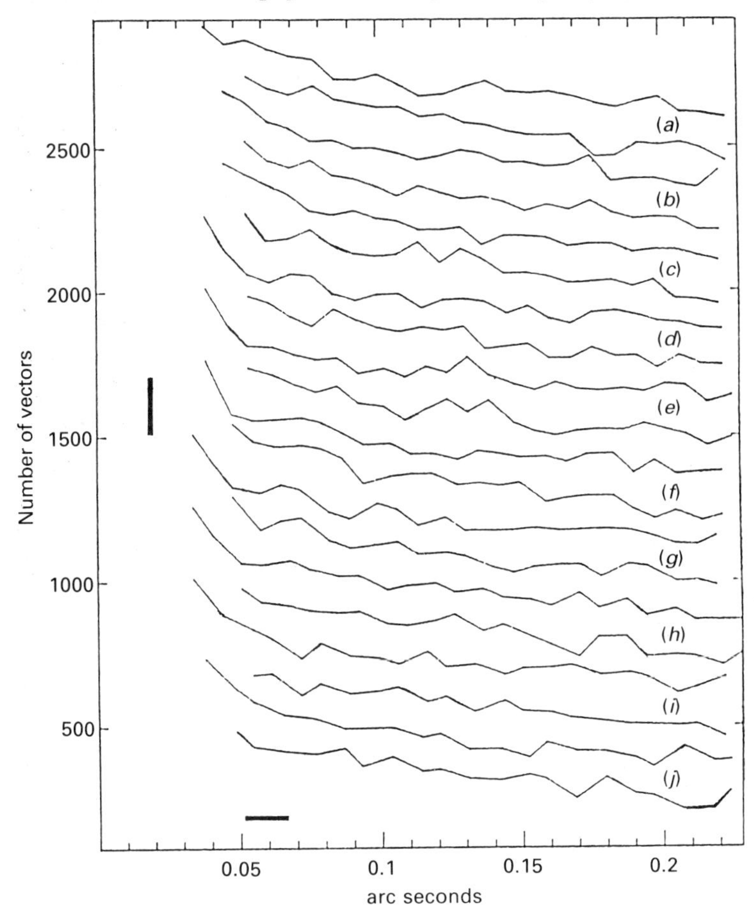

Puzzles from speckles

Optical speckle interferometry was applied to the supernova by astronomers from the Harvard–Smithsonian Center for Astrophysics (Nisenson *et al.* 1987; Papaliolios *et al.* 1988) and from Imperial College London (Meikle, Matcher and Morgan 1987), who carried out the speckle observations at the 4 m telescopes at CTIO in Chile and at AAO in Coonabarabran, respectively.

Their main aim was to determine the supernova's angular diameter as it expanded. Since the supernova is in a galaxy, the LMC, whose distance is rather well determined (accurate to 10–20%), the angular diameter of the supernova can be converted to its actual size in, say, kilometres. This can then be compared to the size as determined from the amount of radiation from the supernova, assuming that it is a black body of a certain temperature, and from the expansion speed and the time since the explosion. The measurements show a very nice progression in time, with a linear expansion over a year at the rate of 2850 km/s (Figure 67). This is a different speed from the spectroscopic

Fig. 67. The size of SN 1987A was measured by the Center for Astrophysics group (filled circles) using the 4 m telescope at CTIO and by the Imperial College group (open circles) using the AAT. The size as viewed through an H-alpha filter increases at a mean speed of 2850 km/s. Courtesy of M. Karovska (CfA).

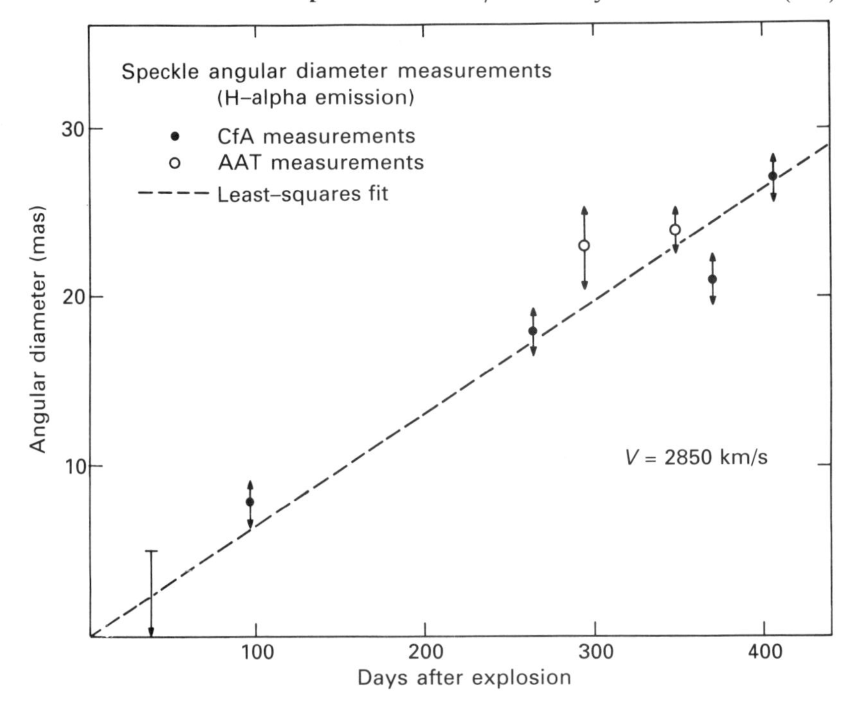

Table 25. *Diameter (thousandths of an arc second) of the LMC supernova*

Wavelength (Å)	Diameter on 1987 April 2 $t=38$ d	Diameter on 1987 June 1 $t=98$ d
Size from speckle interferometry		
3869	<24[a]	
4500	12[b]	23
5330	11	18
6560	1	8
7750	—	15
Size from photometry and black body assumption [c]		
all	2.4	3.4
Size from P Cygni profile[d]		
6563	16	23

[a] Meikle *et al.* (1987).
[b] Papaliolios *et al.* (1988).
[c] Menzies *et al.* (1988).
[d] Assuming constant expansion speed at the speed determined from the trough of the P Cygni profile of H-alpha on that day.

determination of expansion rate by looking at the P Cygni profile and the size of the supernova in H-alpha is different from the sizes determined from other methods (Table 25) which vary a great deal with wavelength. The speckle results are a bit of a puzzle; in particular it is not clear why the supernova's diameter in a waveband which contains H-alpha (6563 Å) should be ten times less than the diameter in other wavebands – everybody would expect the emission line to come from an object which is bigger than and surrounds the photosphere of the supernova.

It is actually a little simple minded to think of 'the' diameter of a supernova. The depth to which we see depends on the wavelength that we observe in. The outer parts get less dense as they expand, and we can see further and further inside the star to completely different layers. The concept of the size of the supernova needs some careful examination and the theory needs working out in detail.

The Mystery Spot

The major puzzle which speckle observations of the supernova have left is the nature of the object which has become known as the Mystery Spot, named after a Californian bar.

Table 26. *The Mystery Spot*

Time (d)	Wavelength (Å)	Separation (mas) and position angle (°)	Magnitude difference	SN mag.	Spot mag.
30	6560	59±8 at 194	2.7±0.2	3.3	6.0
38	6560	59±8 at 194	2.7±0.2	3.0	5.7
38	5330	52±7 at 194	3.0±0.2	3.8	6.8
38	4861		>4	5.0	>9
38	4500	52±7 at 194	3.5–4.0	5.4	8.9–9.4
38	4000		>4.0	7.5	>11.5
38	3869		>4	7.8	>12
47	4861		>4	4.5	>8
47	4921		>4	4.3	>8
50	5876		≥4	3.3	>7
50	6585	74±8 at 16 or 196	3	2.7	6
100	4000–7750		≥4	3.1	>7

The two groups of astronomers who completely independently carried out the speckle observations between 1987 March 25 and 1987 April 14 were surprised to find, off to one side of the supernova itself, a star-like spot of about sixth magnitude, 10% of the brightness of the supernova at that time (Plate 8). The Harvard group was first to process their data, announcing their discovery on May 4 (Karovska *et al.* 1987); then they received a confirmatory phone call from Peter Meikle whose group, having reduced their data on their return from Australia to their home laboratory in London, found the same thing (Waldrop 1987). The fact that both groups saw the same object left no doubt in the minds of the two groups that the Mystery Spot was real. Its separation from the supernova was about 0.6″ – just 18 light-days (Table 26).

Obviously this 'star' was not there before the supernova exploded – it was sixth magnitude and the brightest star in the vicinity had been twelfth magnitude. But how could there be two supernovae? The mystery deepened when the Mystery Spot could not be detected on later observing trips, in 1987 June. It seemed to have faded away. However, this was at about the time when the supernova was at maximum brightness, and the Spot may have been lost in the supernova's glare, rather than really having faded. The Spot has not since been reported to have re-appeared in speckle images as the supernova has subsequently faded.

Whatever the Mystery Spot is (or was!) it seems to be connected with the supernova's asymmetry. Polarisation measurements show that the explosion of the supernova was not round, and the long axis of the elliptical image of the supernova seemed to point towards the Mystery Spot. According to speckle observations in 1988 May by the Center for Astrophysics (CfA) group, the image of the supernova itself can be seen to be elliptical with the long axis pointing towards the Mystery Spot and the end of the ellipse towards the Spot being brighter than the end opposite.

The Mystery Spot had appeared at a distance of about 20 light-days from the supernova within a month of the supernova outburst. That meant that whatever had triggered the appearance of the second image had travelled from the supernova at a speed of two thirds the speed of light at least. This closeness to the speed of light meant that the Mystery Spot could not be explained as the result of the outflow of the supernova material, which was travelling at speeds of less than about one tenth the speed of light.

This gave rise to the thought that the Spot had been formed by radiation which had travelled from the supernova and which had then been intercepted and re-processed in some way by some object nearby. Perhaps there was a cloud near to the supernova which had intercepted some of its light and heat and re-emitted it. The cloud would be generating its energy secondhand. Whatever the re-processing mechanism was, it would have to account for the change of the spectrum of the re-emitted light: the Mystery Spot was significantly redder than the supernova.

There was a difficulty with this idea. As seen from the supernova, the Spot subtended an angle of, at most, 2% of a sphere. It was hard to see, therefore, how it could intercept and re-process 10% of the supernova's energy, even if the reprocessing was very efficient, like a fully reflective mirror. Even bearing in mind the possibility that the cloud was re-emitting in April the ultraviolet radiation which the supernova radiated on February 23, in the first hours that the shock wave broke out from Sk − 69 202's surface, there was still not enough energy available to account for the Mystery Spot. That, then, constituted the mystery.

Speckulations

Martin Rees of the Institute of Astronomy at Cambridge was responsible for an idea about the Mystery Spot which seems to work better than the idea that something is re-processing light or ultraviolet radiation from the supernova. Rees (1987) speculated that the supernova occurred in a binary system and had a faint companion which we have not seen yet.

The supernova shell, outrushing in the explosion, was punctured by the companion. The hole in the shell uncorked as the companion moved on in its orbit. A jet of energetic material from inside the supernova shell was let out, and sped across space, at a velocity of close to the speed of light. The energetic material might have arisen from a pulsar created inside the supernova shell by the supernova. All the power of the pulsar would then come out in the jet. Three light-weeks from the supernova, the jet encountered a nearby cloud. The place on the cloud hit by the energetic jet was heated and emitted the radiation which constituted the Mystery Spot.

Later, the hole in the supernova shell filled up, sealing itself and throttled the jet. In addition the jet pushed the cloud backwards, to greater distances. Thus the Mystery Spot faded away.

Whilst Rees' theory concentrated on the effect of the jet on the cloud, it is interesting to think about the fate of the companion star. As the wave of supernova ejecta passed over the star, its surface was heated and peeled back like a rubber glove being pulled off a finger. The star was hidden inside the supernova shell and would not be seen until the shell expanded and became transparent years after the explosion. The companion star would have been in orbit around Sk – 69 202 (whose mass at that time had been reduced by mass loss from 17 to, say, 10 solar masses) but then, after all the ejected material had passed beyond the companion star's orbit, it would have found itself attracted

Fig. 68. A hole is punched in the expanding layer of SN 1978A by its (hypothetical) companion, and an energetic jet from a neutron star escapes through the hole to illuminate a spot on an otherwise hidden cloud. Is this the origin of the Mystery Spot?

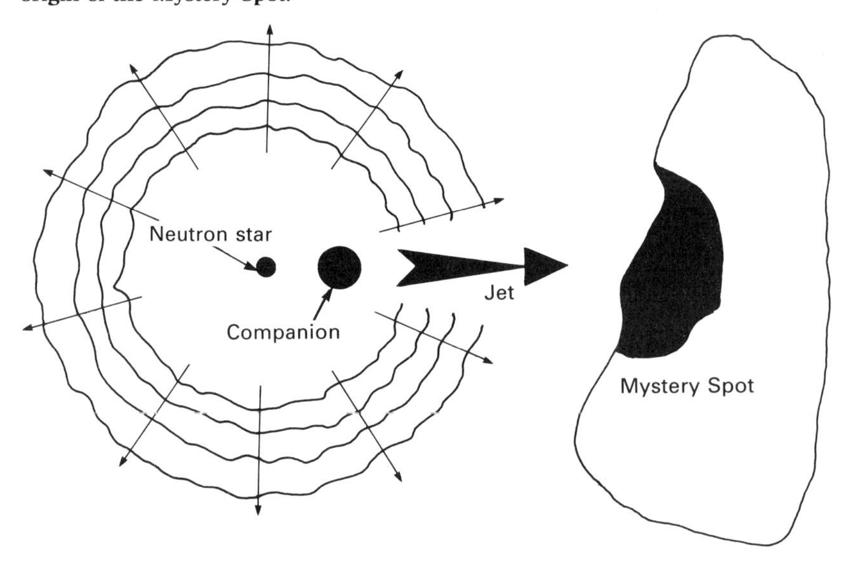

by only the one or two solar mass pulsar left behind. With its gravitational attraction reduced by a factor of 5 or 10, the companion star would no longer be bound in a binary star orbit, and would be released like a sling shot from a catapult. It would set off in orbit around the LMC. Stars released from binary orbits like this are called runaway stars.

This whole sequence of ideas is based on the hypothesis that Sk – 69 202 was originally a binary star. The two stars in the binary system would have orbitted each other at speeds of hundreds of kilometres per second. While it did so, Sk – 69 202 itself would have been emitting its stellar wind. The binary motion would have stirred up the wind to speeds of hundreds of kilometres per second. But IUE observations of the stellar wind, as it emitted narrow ultraviolet spectral lines, showed that the turbulence of the stellar wind (if needed that is where the lines were coming from) was less than 15 km/s. There is an inconsistency here with any theories of Sk – 69 202 which need it to have been a binary star. So it is worth considering other ideas about the Mystery Spot.

Other speculations include the idea that it represented the energy of a coiled spring, which the supernova released like a jack-in-the-box. Some nearby system full of repressed energy was required, able to be triggered by the supernova. No-one has suggested a plausible model of how this concept could be put into practice; the Mystery Spot was not far from being as energetic as the supernova itself and it seems a coincidence that another such energetic event was lying nearby to SN 1987A, waiting to happen.

Another idea was that there was a massive small object, like a black hole, by chance on the line of sight between us and the LMC. The force of gravity around the black hole would bend the light from the supernova, displacing its image like a lens. This constitutes, in fact, a so-called gravitational lens.

Gravitational lenses are known from the study of quasars. In two or three cases a couple of quasars appear near to one another on the sky and with identical spectra – truly a pair of quasars. The total number of quasars known is so small that astronomers cannot believe that this has happened without a good explanation and they conjecture that the pair constitutes the double image of a single quasar. In the most celebrated case, the gravitational lens has been discovered – a galaxy which by chance lies between us and a distant quasar.

By analogy, it may be possible that a similar object distorted the supernova into two images. If it was true that the Mystery Spot – one of the two images – was redder than the supernova itself, perhaps the path of light which formed it passed through some extra interstellar reddening associated with one side of the object.

Calculations showed, however, that the object must be of mass 10^5 solar masses. Black holes this big must form out of something, like a galaxy of stars, a dusty galaxy if the explanation for the difference in colour was accurate. Such a galaxy must be visible if it lies between us and the LMC! It just did not seem credible that a large black hole could lurk completely unseen on the fringes of the LMC and our Galaxy.

Whatever the reason for its disappearance, the Spot is presumably material of some sort, and astronomers are waiting to see what happens when the supernova shell hits it. Parts of the shell were expanding at nearly 30 000 km/s, or one tenth the speed of light. Thus, if light got to the Spot in a month, the shell would get there in something like a year. Astronomers were expecting exciting things to happen early in 1988. As I write in 1988 September there is nothing to report. This has provoked some rebel astronomers in the back rows of conference audiences to mutter sceptical, and probably slanderous, remarks about the speckle observations and the way that they are analysed . . . .

The Mystery Spot remains largely a mystery; in fact without further clues it seems likely always to remain so.

Light echoes

Light echoes, named by analogy with the reflection of sound, were first discovered in a nova over 80 years ago. Up to 20 months after the outburst of Nova Persei 1901, Max Wolf (Heidelberg) and G.W. Ritchey (Yerkes) repeatedly photographed a previously invisible nebula, discovered surrounding the star. The nebula grew in size, up to the apparent size of the Moon. Jacobus Kapteyn realised that the rapid growth of the nebula could not be an ejection of material transported to such a great distance so quickly, but must represent an existing and stationary nebula which had been lit up by the nova outburst.

Henrietta Swope described the only other light echo which had been discovered, surrounding Nova Sagittarii 1936.

Exactly the same as occurred in the two novae happened in the case of SN 1987A. Previously invisible nebulae in the LMC echoed in 1988 March with the light emitted by the supernova when it exploded a year before. These are phantom nebulae, left like the smile of the Cheshire Cat after the supernova has faded.

In 1939, Paul Couderc showed that the reflection nebula illuminated by a pulse of light from a nova lay on the surface of a paraboloid.

A paraboloid is the shape of a surface well known to all amateur astronomers who make their own telescope mirrors. The hollow surface of a parabolic mirror has the property that light from a star at infinity along the main axis of the paraboloid is reflected to a single point, the prime focus. The distance over which each light ray from the star travels in its journey to the prime focus is the same and takes the same time – indeed this is why a paraboloidal reflector focuses. (The scientific name for this is Fermat's Principle.) If the light rays reversed direction and travelled from a flash-bulb at the prime focus of the paraboloidal mirror, then they would travel parallel to the main axis of the paraboloid, in unison 'to infinity'. This is a small scale model of a light echo from the outburst of a supernova embedded in a nebula.

When we look at the light echo from a supernova at a given moment, say one year after the supernova outburst, we are looking at all the places in the nebula in which the journey of light from the supernova to the nebula and then reflected to our eyes has taken an extra year. The lit-up part of the nebula forms a parabolic reflecting surface, with the supernova at the prime focus of the paraboloid.

Fig. 69. A flash of light spreads outwards in all directions from SN 1987A. Some of the light beams direct to Earth. Other light doglegs via reflecting dusty clouds and takes longer. First the flash arrives at Earth by the direct route. Later, the flash arrives via any dusty clouds on the paraboloid. If there are no clouds on the paraboloid except in a sheet of dust lying in front of the supernova, the reflected light forms a ring centred on the supernova. If there are two sheets there are two concentric rings. As time passes, the paraboloid grows and the rings expand.

This diagram greatly exaggerates the width of the mouth of the paraboloid, which is, in fact, very long and thin.

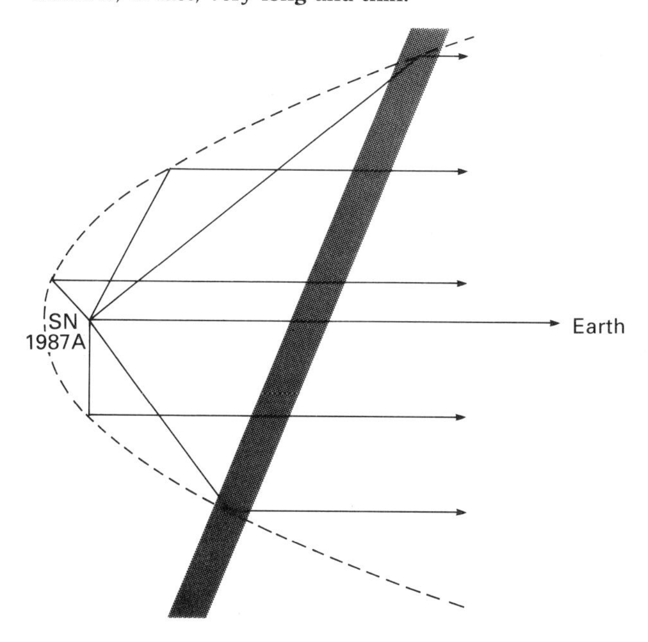

If the supernova is embedded in a vast and complete dusty nebula, then, looking into the paraboloidal nebula/mirror, we should see a disc-like halo of light surrounding the supernova. Couderc's calculations were extended by Schaefer (1987a, b).

In a real situation, the dusty nebula is not complete, because dust in space is very patchy and forms strings and filaments; thus in practice we see not a disc-like halo, but a filamentary structure, where there happens to be dust lying on the paraboloid illuminated by the supernova (Chevalier 1987).

Moreover, for the LMC supernova, its outburst was not a brief flash, but a longer pulse. Thus it is possible for us to see at a given moment nebulae which have been lit by light which originated from the supernova at different times.

Fig. 70(*a*) The region of the supernova, including Sk – 69 202, before it exploded (photograph by David Malin).
(*b*) An image obtained by Suntzeff *et al.* (1988) in which the supernova is hidden behind the central occulting disc, at the intersection of the cross-shaped diffraction spikes (picture courtesy of Ron Olowin).
(*c*) A photograph showing the supernova (photograph by David Malin): the diffraction spikes arise from the supports of the camera inside the telescope and differently oriented in the two telescopes used for (*b*) and (*c*)).

The nebulae in the upper half of the pictures are always present, but in the last two pictures the two circular rings, concentric with the supernova and most prominent towards the centre of the pictures, appeared after the supernova. They are the light echoes reflected off two sheet of dust which lie hundreds of light-years in front of the supernova.

Compare the positions of the rings in (*b*) and (*c*) relative to the patterns of stars and observe how both rings have significantly moved outwards in four months as the light flash propagates outwards in the sheets.

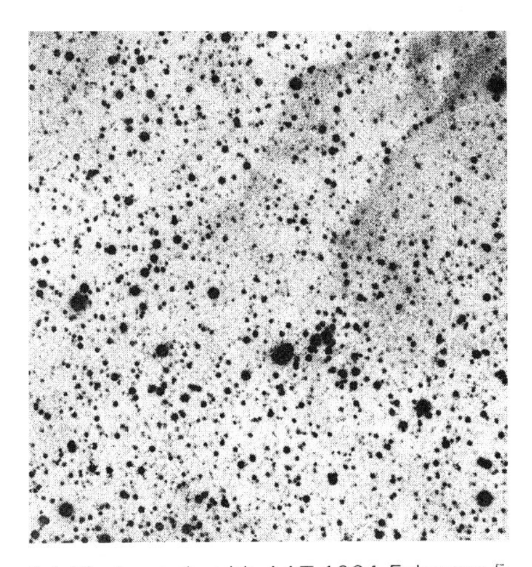

(*a*) Photograph with AAT 1984 February 5

Astronomers using the Carnegie Institution's 1.0 m telescope (by coincidence the telescope was built using a donation by and named for Henrietta Swope, discoverer of the second known light echo) and the ESO's 3.6 m

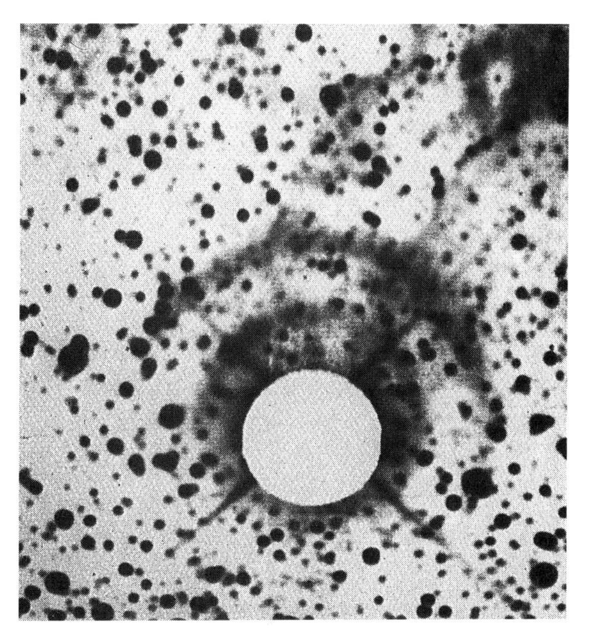

(*b*) CCD image with 1.5 m telescope at CTIO 1988 March 19–2

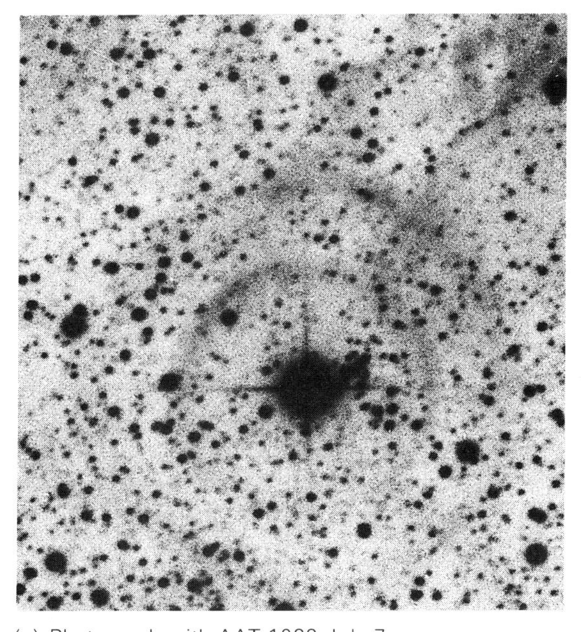

(*c*) Photograph with AAT 1988 July 7

telescope, both in Chile, discovered in late February and early March 1988 the light echoes from SN 1987A. They consisted of two faint, ring-like, reflection nebulae surrounding the image of the supernova. The nebulae had radii of about 30″ and 50″, respectively.

First reported by Crotts (1988) of the McDonald Observatory from exposures in the first week of March, the light echoes were confirmed on two exposures taken 1988 February 13 and March 16 at ESO (Rosa 1988, Guiffres *et al.* 1988). In retrospect they were visible on ESO pictures taken as early as 1987 August 16. Detection of the faint light echoes was technically difficult because of the brightness of the supernova in their midst. Early in 1988 March the supernova had just dropped below visual magnitude 7.0 – easily visible in a pair of binoculars and dazzlingly bright in a large telescope. To avoid flooding their detectors with unwanted scattered light, both the Carnegie Institute and ESO astronomers obstructed the image of the supernova with a dark patch at the focus of their telescopes. Because of the similarity of this device to an eclipse of the Sun by the Moon, which reveals the solar corona, the apparatus which they each used is called a coronagraph.

The light echoes from SN 1987A did not look like haloes – they were rings, rather than discs. This was because the dust did not completely surround the supernova, but lay in two sheets in front of it. The two rings lay at the intersection of the paraboloidal surface and the two sheets. Since the nebulae were nearly circular in shape, the sheets lay across the line of sight.

Suppose that a dust sheet lies a distance d light-years in front of the supernova, S. Light which has passed directly from the supernova through the dust at the point P, to the Earth, reached us in 1987 February. Light which radiated at an angle A has taken an extra time t (one year in this case) to dogleg via the dust sheet at the point D to us, and is lighting up a circle on the dust sheet of radius r light-years. D lies on a paraboloid whose focus is at the supernova – the paraboloidal surface lies a distance $0.5t$ behind the supernova, so any dust there is lit up at the same time as D. The dust sheet lies across the line of sight, so DPS is nearly a right angle triangle and Pythagoras' theorem gives:

$$(d+t)^2 = d^2 + r^2.$$

So

$$d = (r^2 - t^2)/2t.$$

With $r = 30$ light-years and $t = 1$ light-year, $d = 450$ light-years. The second ring has a radius of 45 light-years, and t is still 1 light-year, so it lies on a second sheet about 1000 light-years in front of the supernova. As t increases, D moves outwards and r increases: the rings get larger.

Fig. 71. Figure 70(*a*) has been subtracted by David Malin from Figure 70(*c*) in order to display the difference. Apart from the slightly different sizes of the stars in the two photographs, the major differences are the image of the supernova itself, with its diffraction spikes, and the two light echoes. AAO photograph.

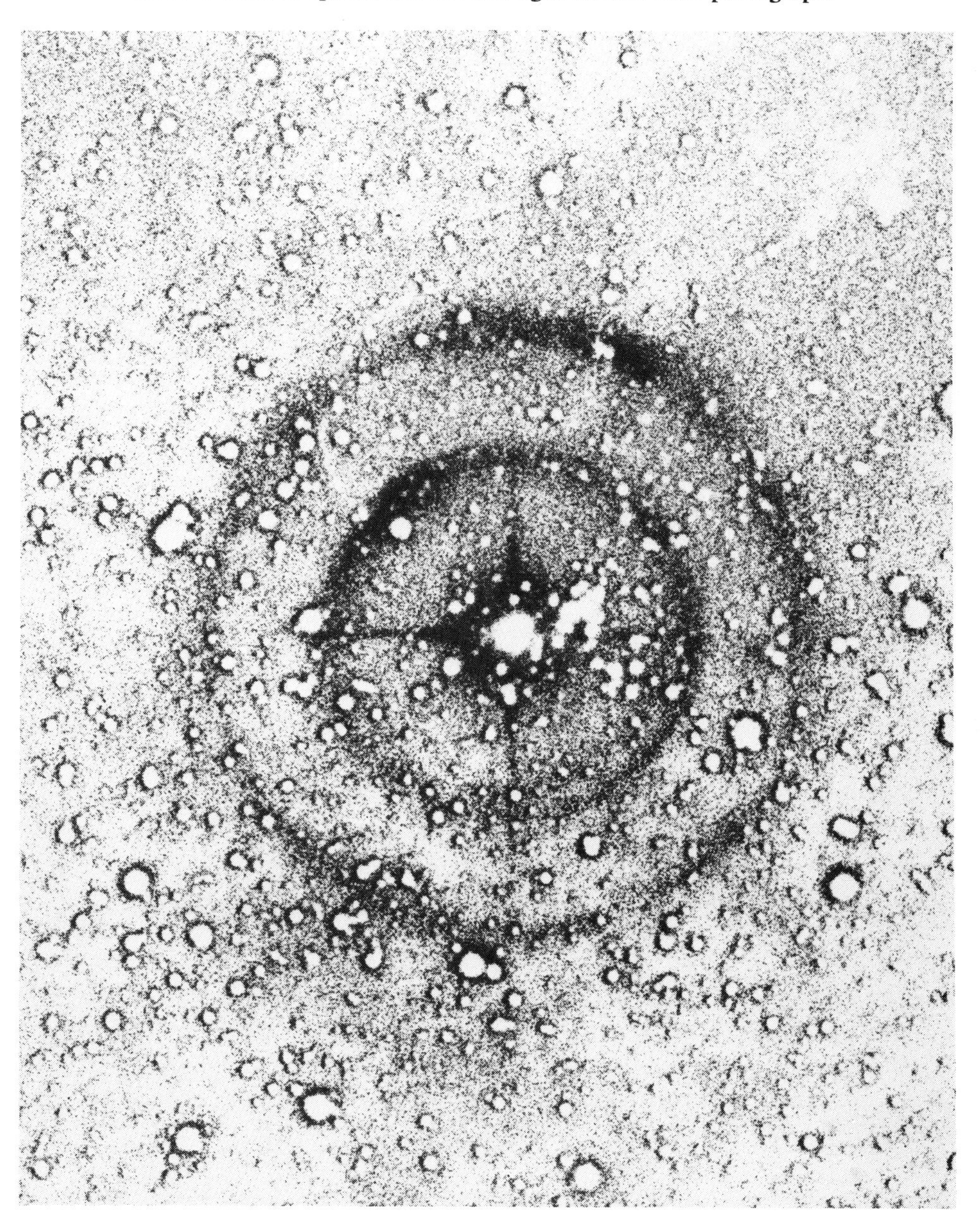

Crotts (1988) calculated that the two sheets lie 450 and 1000 light-years in front of the supernova. Thus they were nothing particularly to do with the supernova or the star, Sk – 69 202, which exploded; they were in the foreground of the LMC and lit up by chance. The paraboloidal surface representing the locus of the parts of space lit up by the supernova is very long and thin, almost tubular with its open end pointing to Earth.

Dust in the two sheets which lies directly between Earth and the supernova would have been illuminated by the light passing directly from the supernova to

Fig. 72(*a*) The spectra of the two light echoes in 1988 March (top two panels) did not at all resemble the spectrum of the supernova at that time (bottom panel). (*b*) The spectrum of the outer echo in 1988 March, particularly the H-beta dip and bump near 4860 Å, resembles the spectrum of the supernova 11 months

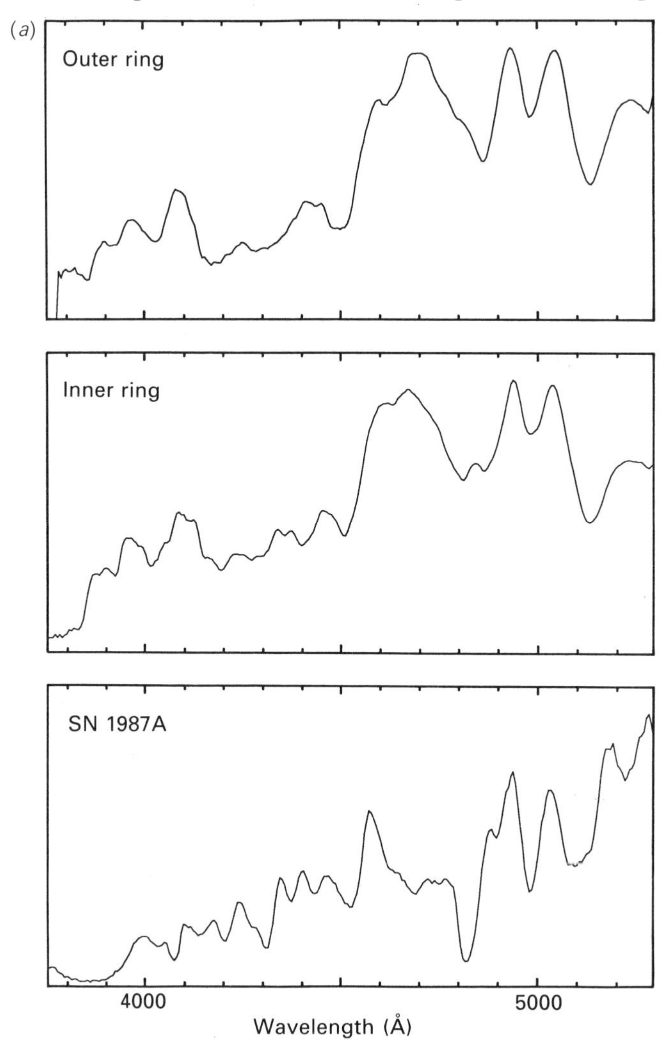

us; at the time when the light from the supernova reached us, the light echoes would have lit up a small patch of nebula immediately superimposed upon the image of the supernova. The echoes have grown outwards like ripples ever since, the open mouth of the paraboloidal tube opening wider.

The nebulae lit by the light echoes became fainter as the supernova's light became more diluted by travelling so far. The ESO exposures confirmed Crott's calculation that the ring's radii at their discovery should be expanding by 1″ and 3″ per month as the pulse of light from the supernova rippled outwards

earlier, in 1987 April (central panel); the spectrum of the inner echo is similar to the spectrum of the supernova 10.5 months earlier, between 1987 April and May (top two panels). Suntzeff *et al.* (1988); diagrams courtesy of Ron Olowin (St Mary's College of California).

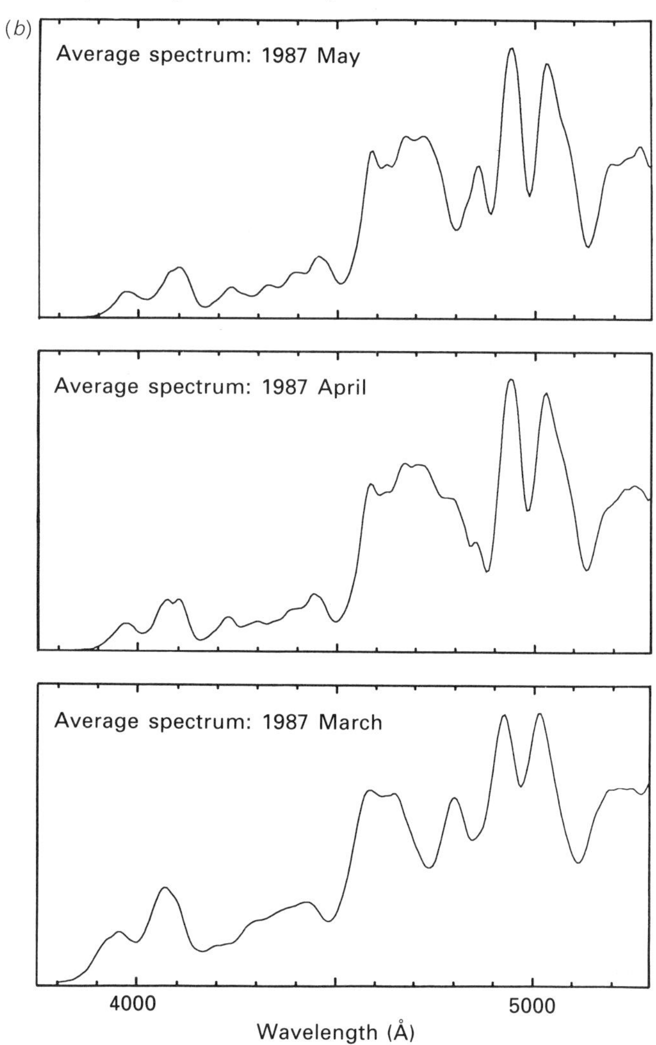

through dust which reflected the light echoes.

One arc second at the 200 000 light-year distance of the LMC is approximately 1 light-year, and so, when they were discovered, the rings were expanding by 1–3 light-years per month. This is up to 30 times greater than the speed of light, which is 1 light-year per year (by definition). The expansion of the rings is thus 'superluminal' – faster than the speed of light.

This does not violate the theory of relativity, because nothing material is actually moving faster than the speed of light. The dust is stationary – it is the way the dust is illuminated which makes the nebula appear to expand so quickly. (If a lighthouse beam sweeps rapidly across a distant cliff face, the illuminated patch might travel faster than the speed of light across the rocks, but the rocks are not moving – so, no problem with the theory of relativity!)

Of course the real situation is more complicated than this idealised model. According to Guiffres *et al.* (1988) and according to observations with the 4 m telescope at CTIO, (Huchra and Olowin 1988; Suntzeff *et al.* 1988), the light echo on part of the inner ring had a spectrum at that time with the same appearance as the spectra of the supernova in 1987 May. This ring seemed, therefore, to be the echo of the supernova when it was at its brightest. However, the spectrum of the outer ring was the same as the supernova had in 1987 April, *before* it reached maximum. Not all of each ring need be an echo from the same moment in the supernova's outburst; the outer echo travelled 11 light-months further than the light which we see from SN 1987A and the inner echo about 10 light-months. The rings are braided plaits of dust, the strands picking up echoes of different moments in the supernova's light curve – a 'light reverberation' in fact.

Additional light echoes from isolated clouds in the LMC and from material behind the Supernova appeared early in 1989.

Since the ultraviolet light from the supernova flashed rather briefly in the very first days after the start of the explosion, there should be an 'ultraviolet echo' lying outside the light echoes. The IUE satellite has looked for such an ultraviolet echo, with no certain detection (Panagia, 1988).

12
Neutron star or black hole?

As SN 1987A gets larger and larger it gets less and less dense, and so more and more transparent.

In everyday life we know that material is transparent when it forms a film of surface density less than some number of grams per square centimetre across our vision. Above this critical surface density it becomes opaque. For example, glass is murky if it is thicker than a few centimetres – say 10 cm. In this case its surface density is 50 g/cm^2. Wood or metal is ordinarily not transparent if it is 10 cm thick – but shaved thin to a surface density of 0.1 g/cm^2 (wood) or 0.01 g/cm^2 (metal) it is transparent. The volume density of air is 0.0013 g/cm^3 and a column, say, 10 km long diminishes the visibility of things in the distance – thus air is becoming opaque at 1 kg/cm^2.

Similarly we know that our flesh is transparent to X-rays, but our bones tend not to be – and we have a projected surface density of about 5 g/cm^2.

For the fully ionised material of a star the opacity is caused by the interaction of light with the electrons in the plasma in the body of the star. The cross-section of an electron for such an interaction is called the Thomson cross-section and it is equal to 0.7×10^{-24} cm^2. If more than 1.3×10^{24} electrons fit over a square centimetre, their cross-sections total over 1 cm^2, and the material is opaque.

If the star material is helium these 1.3×10^{24} electrons per square centimetre are associated with $0.5 \times 1.3 \times 10^{24}$ helium nuclei because two electrons are associated with each nucleus. Avogadro's number of helium nuclei (6×10^{23} of them) weigh 4 g. Thus the star material is opaque if it has a surface density of 4 g/cm^2 or more.

The supernova material is of mass 15 solar masses, i.e. 3×10^{34} g. When it has expanded to a radius R cm its surface density is on average $3 \times 10^{34}/\pi R^2$ g/cm^2. This is at the critical value of 4 when $R = 10^{17}$ cm, i.e. 0.1 light-year. Since the supernova is expanding at a rate of 0.1 of the speed of light, it reaches 0.1 light-year in one year.

The calculation is crude and should not be taken exactly, but indicates that the supernova would become transparent in 1988–9. This was, in fact, the time when the radioactive cobalt at the centre of the supernova became readily visible in gamma rays and infrared radiation.

In consequence of the reduction in density through the supernova we would see further and further into its inside. Eventually, we would be able to see the object that formed at the centre of the supernova when the core of Sk – 69 202 collapsed. Presumably this would be a neutron star – or perhaps a black hole.

Neutron stars

Neutron stars are objects which have the mass of a star and which are made of neutrons. They are typically 10 km in radius and 1 solar mass. Like ordinary stars, and as explained in Chapter 5, they remain in equilibrium by balancing two forces – their own force of gravity, which pulls their material inwards, and a pressure force which pushes outwards. The difference between neutron stars and other stars is in the way that the pressure force, P, is related to density, ρ and temperature, T – the equation of state. The pressure force is different from the ordinary gas pressure expressed by Boyle's law:

$$P \propto \rho T.$$

If the temperature of a neutron star is low (this means lower than some millions of degrees), the pressure in neutron material has no temperature dependency and the equation of state is either

$$P \propto \rho^{\frac{5}{3}}$$

for lower density material or

$$P \propto \rho^{\frac{4}{3}}$$

for the higher density material. Because the power of ρ is smaller for the higher density material's equation of state, the higher density material can be more easily compressed.

These equations of state arise because of the quantum mechanical nature of neutrons: the material is called degenerate because of the lack of temperature dependency. The equations of state give rise to one of the most puzzling properties of neutron stars – the relationship between their size and mass, M.

The mass–radius relationship for neutron stars is quite different from the relationship for Main Sequence stars. Main Sequence stars have the quite reasonable property that the more massive stars are bigger in size. But neutron stars have a size which is proportional to $M^{-\frac{1}{3}}$ – the more massive stars are smaller.

Suppose that I make a neutron star and add more neutrons to it. As I add more neutrons the star gets more massive. Because of the extraordinary mass–radius relationship, it therefore gets smaller. Its density gets larger (because there is more material in a smaller volume). It becomes relatively easier to compress the star further, and the next amount of neutrons which I add makes the neutron star much smaller. The result is that as I make the neutron star approach a certain limiting mass, the neutron star gets vanishingly small.

There is thus a certain limiting value to the mass of a neutron star; it is about 1.5 solar masses and is called the 'Chandrasekhar limit'. It is a true limit only for 'cold' neutron stars, but, since all stars eventually cool, it is an inevitable limit.

If a neutron star forms with mass above this limit, then it will collapse to a very small size.

If a neutron star forms with mass below this limit but more mass is added to take it over the limit, then too it will collapse. In fact, white dwarf stars have exactly the same property as neutron stars: the pressure is from electrons rather than neutrons, but electrons have the same quantum mechanical properties as neutrons (namely, they have spin $\frac{1}{2}$) and they have the same degenerate equations of state. Because of this property, if a white dwarf forms in a binary system and accretes matter to drive it over the Chandrasekhar limiting mass, it collapses. The result is a Type I supernova.

Pulsars

Until the discovery of pulsars in 1967, neutron stars were a theoretical concept thought not to exist in reality. But the discovery of pulsars by Nobel prizewinner Antony Hewish and his graduate student Jocelyn Bell showed that they do, in fact, occur in nature, produced by supernovae.

The name 'pulsar' is an acronym for 'pulsating radio star'. The pulsations take the form of very regular brief flashes of radio emission. This was the property which led to their discovery.

A pulsar is a rotating neutron star, formed from the collapse of the core of a progenitor star. The neutron star material has a very high conductivity and trapped within it in the collapse is the star's original magnetic field compressed to a very high intensity – perhaps as high as 10^{16} G.

The collapse of the core of the star is radially inwards, and angular momentum is, at least to some degree, conserved. The core of the star may have been slowly rotating, say once per month like our own Sun. When its size is suddenly reduced, there is a great reduction in the moment of inertia of the star. It increases its angular frequency to compensate.

The angular frequency before and after the collapse are related by the following equation which expresses the conservation of angular momentum of an object of mass M, rotating with period P, during its collapse from radius R to radius r, and which keeps the same shape (expressed by the fact that k in the equation remains the same on both sides):

$$(Mk^2R^2)/P=(Mk^2r^2)/p.$$

The new period p is given by:

$$p=P(r/R)^2.$$

If $r=10$ km and $R=6000$ km, and $P=1$ month ($=3$ million seconds), then $p=10$ s.

What was a slow rotation speed of, say, one rotation per month becomes a fast rotation, with a period of seconds.

The neutron star is rotating and contains a magnetic field. The field is not necessarily aligned with the rotation, but lies across the rotation axis, in the same way that the Earth's magnetic pole lies offset by an angle to the pole of rotation. The rotation of a magnetic field is similar to an electric generator, and the same consequences occur. Electrical currents develop from the surface, pouring relativistic electrons (and protons) along the neutron star's magnetic field into the region around the neutron star.

The flow of the electric currents produces sychrotron radiation. The radiation is in the form of radio waves (light and X-rays too). The radiation flows in beams like a lighthouse. The beams are fixed to the magnetic field and are carried around by the neutron star's rotation – hence the pulsars appear as flashes of radio emission. Moreover, the electrons which escape from the neutron star's influence produce a general background of synchrotron emission. They create a white nebula, called a plerion, centred upon the pulsar.

The Crab Nebula is the remains of a supernova like SN 1987A but which exploded in our Galaxy in the year 1054. It contains all these phenomena. A pulsar at its centre rotates 30 times per second at the present time, having slowed over the last thousand years as its energy has radiated away. The nebula itself is a network of filaments which represent the shattered remains of the body of the star which exploded, and the network encloses the electrons which have escaped from the neutron star. These electrons radiate synchrotron radiation and can be seen as a radio and white light nebula lying within the filaments.

A supernova of the more distant past, which was not recorded by history, but whose remains can be seen in the constellation of Vela, also contains a pulsar rotating 11 times per second, and residual synchrotron radiation.

The supernova remnant called 3C58 does not contain a pulsar which we can see directly – perhaps its lighthouse beams never sweep across the direction towards us – but 3C58 does contain a bright synchrotron nebula and a pulsar is inferred to exist inside it, generating the electrons which produce the synchrotron emission.

If SN 1987A produced a neutron star, then eventually phenomena like this would become visible at the site of the explosion. According to Martin Rees (Royal Astronomical Society Discussion Meeting on Pulsars, 1987 November 13) the signature of a pulsar's activity would be (*a*) fast particles, (*b*) synchrotron radiation and (*c*) pulsed emission, the total luminosity of $L = 10^{40}$ erg/s (for a rotation frequency of 100 times per second) creating a bubble of plasma which would inflate inside the supernova's expanding envelope (Table 27). If it spins fast, this energy will become visible from the supernova's envelope, as the radioactivity from the nickel and cobalt fades away.

Calculations of when everything becomes visible depend on whether the supernova envelope fragments into individual cloudlets or filaments and whether it has holes or thin patches because it is non-spherical. Unpulsed radiation from the pulsar's surface (X-rays from its cooling) will probably not be visible for more than ten years after the supernova's explosion, but pulsed radiation from the pulsar's lighthouse-type beam could well be seen before that. However, we would only be able to see the pulses if we on Earth were in the sweep of the pulsar's beam.

The probability of being in the beam of the pulsar depends whether the beam is thin and narrow like a pencil or thin and fan-like. If pulsars have pencil-like beams, the probability (for an arbitrary orientation of the pulsar's spin axis relative to us) that we are in the sweep of the beam is about 1 in 5; if the beam is fan-like then the probability that we are in the sweep of the beam is much better.

X-ray pulses could not be visible for a couple of years, but optical pulses could appear any time after about a year; radio pulses would find it more difficult to escape. If we were not in the sweep of the pulsar beam, we might be able to see scattered radiation from the supernova material nearby. The pulses would not be individually visible because of time delays as the pulses echo from different bits, just as individual notes of music from a tape recorder held in an underground railway tunnel blur into an annoying general rumble.

Table 27. *A neutron star in SN 1987A?*

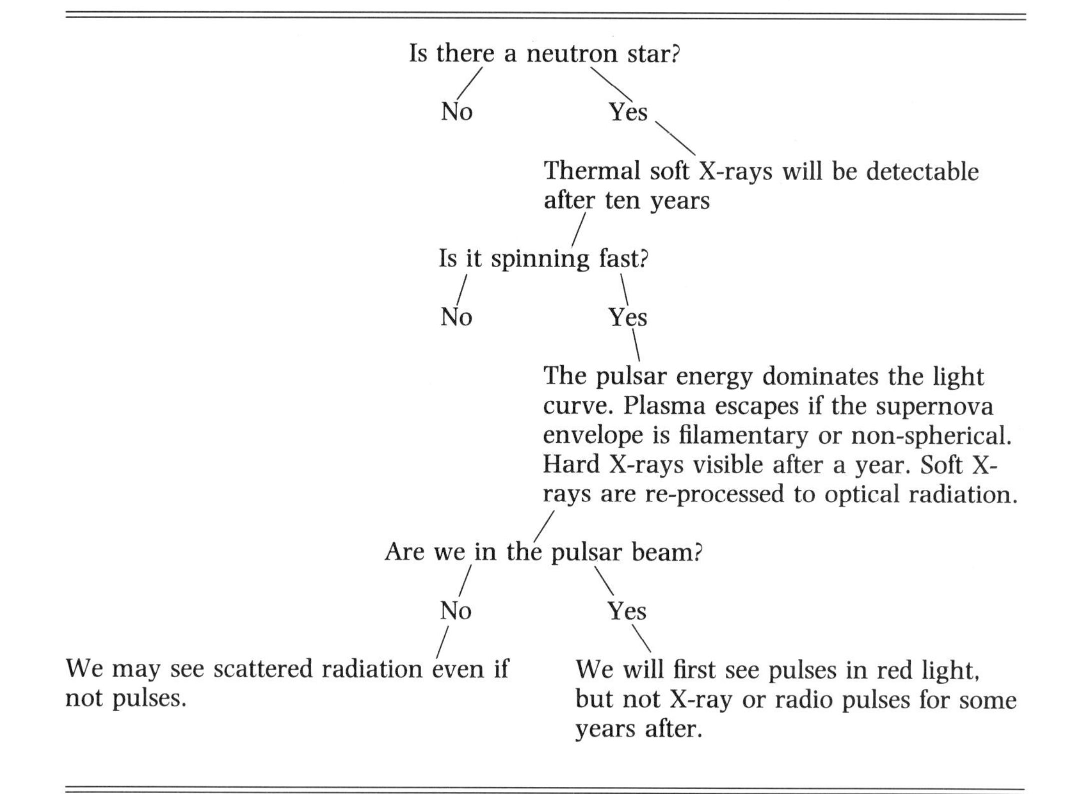

Black holes

Not all supernovae produce neutron stars. We know this because not all supernova remnants are like the Crab Nebula or 3C58. Tycho's supernova of 1572 and Kepler's of 1604, for example, have formed remnants which have spherical shells of filaments which contain nothing – no pulsar, no synchrotron emission. These supernovae were, however, Type I supernovae, as judged from their light curves determined from the careful observations gathered by Tycho and Kepler themselves. It is thought that Type I supernovae cause the complete disintegration of their progenitor stars, so it is not surprising that their remnants contain no neutron star/pulsar.

There are, however, recently formed remnants of Type II supernovae which do not contain pulsars or synchrotron emission. The famous example is called

Cassiopeia A. No supernova was definitely recorded in this region[†], but from the speed of outflow of the filaments it must have occurred in 1670 or thereabouts. The filaments have the composition which is expected if they had belonged to a massive star whose disintegrated interior was made of oxygen produced by carbon burning. The supernova must have been formed from the explosion of a massive star and must therefore have been of Type II.

Since it is empty of a pulsar but was formed by a core collapse, Cassiopeia A probably contains a black hole.

A black hole is a star in which the internal pressure force has been completely overcome by the force of gravity, shrinking the star to a very small size. The force of gravity at the surface of such a star is so strong that nothing can escape from its surface. Anything which tries to leave is pulled back by the gravitational force.

This is not such a very strange idea when you think about it. On Earth we are accustomed to the idea that a ball thrown upwards returns to Earth, pulled back by the force of gravity. A ball thrown up at a faster speed flies higher before falling back. If thrown fast enough, like a space rocket, the ball will escape from the Earth's pull, because the Earth's force of gravity is weak. If the Earth's force of gravity could be increased, then it would be necessary to throw the ball faster for it to escape. If the force of gravity could be made so strong that it would be necessary to throw the ball faster than the speed of light for it to escape, then escaping would be impossible, because nothing can travel faster than the speed of light. Nothing material could leave the Earth if its gravity was this strong. The Earth would have become a black hole.

Black holes have the additional rather surprising property that not even light can leave them. The reason for this is that, according to the general theory of relativity, not only material things but photons, too, lose energy as they travel against a gravitational field. They may lose all their energy if the field is strong enough. Thus there may be a star with such an immense gravitational force that no light pulse can leave it. No information that the star exists can leave it. The star is black. It is a hole, in that anything dropped in never gets out.

There are extraordinary paradoxes which are generated by a strong gravitational field, and black holes have some weird and wonderful properties. I have no need to go into these paradoxes. For my purpose a black hole is no more than a collapsed stellar core; its gravitational force is so large that it has conquered the internal pressure and squashed the star to a very small size.

† Perhaps the first Astronomer Royal, John Flamsteed, observed the supernova. He recorded a faint star, which he called 3 Cassiopeiae, in the region of Cassiopeia A. The star has since disappeared, as a supernova would have been expected to do.

Did SN 1987A form a black hole?

One clue as to whether SN 1987A contains a neutron star or a black hole is the suggestion that there may have been two neutrino bursts – one burst observed by the Mt Blanc neutrino observatory five hours before the burst observed by Kamiokande and IMB (see Chapter 8). And this may indicate that SN 1987A produced both a neutron star and a black hole, first one and then the other.

Some scientists believe that the early neutrino burst detected by the Mt Blanc neutrino observatory was spurious – an unlucky mistake. For them the only significant neutrino burst is the one detected by IMB/Kamioka, and it signalled the production of a neutron star: that is what we will eventually see when the supernova becomes transparent.

But those scientists who continue to believe the Mt Blanc detection of a neutrino burst as well as the IMB/Kamioka burst take the view that two neutrino bursts have to be explained. Although it may have been that Mt Blanc saw an unrelated event by coincidence in the hours before the supernova was discovered, we do not get many opportunities to examine the consequences which follow from seeing two neutrino bursts on the same day.† (On the other hand, Koshiba, at the George Mason Workshop, dismissed the Mt Blanc result with the words, 'The credibility of an experimental result should not depend on weak supporting experimental evidence, but nor should it depend on appeal to unsubstantiated theories.')

What, then, are the theories (whether thought to be unsubstantiated or not) which are proposed to account not only for the formation of a neutron star, but something else too, and to explain the five hour separation of the two events. Was the second burst the formation of a black hole?

The scenario for how this could happen is that SN 1987A produced a neutron star whose mass was above the Chandrasekhar limit and just too much for it to survive.

Suppose that the collapse of a stellar core produces a neutron star whose mass is above the limit. The star will be 'hot' when it is first formed. It has some extra pressure force within itself which, at first, enables it to resist the additional gravity of the additional mass.

† In one of her books Agatha Christie puts wise advice for scientists into the mouth of her detective heroine: 'Any coincidence', said Miss Marple to herself, 'is always worth noticing. You can throw it away later if it *is* only a coincidence.'

Such a star will not collapse further for some time. But after the star cools the neutron star finds itself in a position where it cannot resist further collapse to a vanishingly small size. The only possible outcome is a black hole.

An alternative or additional explanation for the delay in collapsing is that the neutron star is rotating quickly so that centrifugal force effectively reduces the gravitational force of the additional mass. But then it radiates its kinetic energy through the electromagnetic waves which it sends into the surrounding space (or by emitting other kinds of radiation). Its rotation slows and the centrifugal force no longer contributes to keeping the star up against gravity. This also makes the star collapse to a black hole.

Thus it is possible that the two neutrino bursts in SN 1987A represent two stages in the collapse of a stellar core – first, to a neutron star, second, to a black hole, five hours later (Hillebrandt *et al.* 1987). Perhaps SN 1987A was not on a direct route to make a black hole, but a two sector flight with a five hour stopover.

Collapse, pursuit and plunge

An eccentric view (Stella and Treves 1987) of the two neutrino bursts, but an idea which is provocative and very interesting, is that perhaps the supernova initially produced two neutron stars which merged to form a single one.

Why should a core collapse have produced two neutron stars?

We have seen earlier (Chapter 9) that the supernova envelope was not round, and the existence of the Mystery Spot (Chapter 11) also implied that there was an axis of some sort in the outflow of the material of the supernova. The outflow was not symmetric if there is a phenomenon which implies that there was an axis.

What does this mean for the core collapse? Presumably, if the outflow of envelope material was not symmetric, neither was the infall of the core. If the core did not contract uniformly, retaining its sphericity, then it may have either pancaked into a disc-like object by falling faster around its equator.

In either case it might be that more than one object would form, separated each from the others, rather than one object at the centre of the core. Perhaps there were two objects formed, such as two neutron stars, say (or one neutron star and a black hole). According to this theory, the simultaneous formation of the two neutron stars caused the first burst of neutrinos, the one seen by the Mt Blanc Observatory.

The neutron stars would remain for a time in orbit, circling each other, very close and with a very short period – for concreteness Stella and Treves (1987) envisage an orbital separation of 590 km and an orbital period of 0.2 s. Obviously this is an ultra short period binary, considering that the fastest binary known has a period of 11 minutes (the star is an X-ray source called EXO 1820–30, and consists of a neutron star and, perhaps, a white dwarf type of star). But the idea is a straightforward extrapolation of what is already known.

The orbit of the two neutron stars formed by the collapse of SN 1987A decayed, and the two neutron stars coalesced. In the merger a further burst of neutrinos was emitted.

The sequence of events described is called the 'collapse, pursuit and plunge' scenario.

The theory is built on the concept of gravitational radiation as the mechanism which causes the orbitting neutron stars to lose energy. Gravitational radiation is a theoretical concept up to the present time, since none has ever been detected.

The theory of gravitational radiation is based upon Einstein's General Theory of Relativity. Consider two test particles sitting side by side in space near to an object from which the gravitational field is varying. A binary star is one example of an object which will produce a varying gravitational field. The attraction of the pair of particles by the binary star varies with the orientation of the binary star. The test particles are pushed together and pulled apart in synchronism with the binary star period.

You can imagine a gravity wave from the binary star passing across the pair of test particles; they are affected by it like corks bobbing on the sea. The wave propagates into space. It carries off energy from the binary star. The separation of the two stars must diminish as energy is carried off; they approach each other and their period gets smaller.

Although gravitational radiation has never been detected it is believed that its effects have been seen in just such a case as described. A star called the Binary Pulsar consists of a pulsar orbitting another small star. Because the pulsar is a regular clock, its position in its orbit can be very accurately mapped by the effect of its position on the arrival time of the pulses: pulses emitted from the far side of the orbit arrive at Earth behind schedule, and pulses from the pulsar when it is on the near side arrive earlier. The orbit of the pulsar is very accurately described by the General Theory of Relativity, including the effects of gravitational radiation.

The amount of gravitational radiation which two neutron stars formed in a supernova explosion give off increases as they approach each other, and they approach each other faster still. Eventually they coalesce. Stella and Treves calculate that coalescence occurs some hours after the formation of the binary star system. At the time of the merger, the stars are 30 km apart and their orbital period is 2.4 ms.

Although such a thing has never been seen to happen, it is believed that mergers of binary stars happen quite frequently in the galaxies nearby. There are detectors existing and under construction that are intended to measure the gravitational waves given out as binary stars merge. The 'strain' in space produced by such an event in a nearby galaxy is estimated to be 1 part in 10^{22}. This means that the distance between two test particles (say, freely suspended pendula) which are 1 km apart must be measured to better than the diameter of a proton. While the technology does not yet exist for this, it is within the horizon of present-day engineering.

There are less sensitive gravitational wave detectors in existence at the present time. If the core of Sk − 69 202 collapsed spherically, then no gravitational waves were emitted, since the gravitational force outside a spherical object is the same no matter what size it is. But if the core collapsed in a pancake shape then some gravitational waves were emitted. If a binary neutron star was formed which then coalesced, it too should emit gravitational waves as explained above.

Gravitational waves travel at the speed of light, and so too, if their mass was zero did the neutrinos from SN 1987A. Thus if any gravitational waves were emitted from SN 1987A, they must have arrived at Earth at the same time as the neutrinos. Unfortunately only the world's two least sensitive gravitational wave detectors (in Rome and in Maryland) were in operation at the time of the Mt Blanc burst and none was operating well at the time of the IMB/Kamioka neutrino burst (Amaldi *et al.* 1988).

There is a suspicion that, in both the gravitational wave detectors, pulses were seen at the time of arrival of the Mt Blanc neutrinos; the probability that this is a coincidence has been calculated at three chances in a million. Most astronomers regard the claim with scepticism, and think that nothing really significant was seen.

One of the barriers against believing that gravitational waves from the LMC supernova have been detected is that, according to the usually accepted theory of gravitational waves, the energy released in the LMC to make observed gravitational wave pulses would have to be vast – equivalent to the total annihilation of 2000 solar masses of material. It is hard to see how this can be released from a star of 17 solar masses! It would be necessary to abandon the accepted theory of gravitational waves. John Weber, in charge of the Maryland gravitational wave detector believes it is necessary (Weber 1988). Only then can the claim to have detected gravitational waves from the LMC supernova be right.

There is nothing in principle wrong with the claim, except that its credibility depends on a difficult-to-accept neutrino burst, individually insignificant gravitational wave detections and a new theory of gravitational waves; astronomers are just not prepared for all of this at once.

Did SN 1987A form a neutron star?

There is very little evidence that SN 1987A produced a black hole, or a binary neutron star.

The straightforward view is that SN 1987A produced a single neutron star. The calculations of the energy and the spectrum of the neutrino burst produced by the collapse of the core of Sk − 69 202 to a neutron star are in very good agreement with the observations made by the Kamioka and IMB neutrino observatories (Burrows 1988).

It was, however, worrying that no neutron star or pulsar-like phenomena had, by the end of 1988, made themselves perceptible. No matter what form energy took as it was released by the pulsar, it would have to feed the total power radiated by the supernova (Table 27). But from 1987 July until the end of 1988, the supernova's radiated power – light + infrared + X-rays + gamma-rays – showed a drop-off with time which was exponential in form, and exactly corresponded to the release of radioactive energy from ^{56}Co. There was little room for any additional source of power, such as a pulsar, and Arnett (1988) estimated that no more than 5% of the supernova's power came from one. Any pulsar in SN 1987A could only be as powerful as, or less powerful than, the Crab pulsar. This is surprising, since the Crab pulsar has had nearly 1000 years to lose its energy and to decrease its radiated power. If a pulsar exists in SN 1987A it was born slow, or without a strong magnetic field, and is an inefficient generator and a weak source of electrons, with all that that implies.

Although most of the evidence available in 1988 was about the existence of a neutron star in the supernova was negative, there was an intriguing result which pointed to one's existence – the X-ray and gamma-ray flare of January 1988.

The X-ray 'light' curve as observed by the Ginga satellite showed a sharp peak around 330 days after the supernova explosion – it was particularly pronounced in the softer energy X-rays seen by Ginga. The flare was the shape of and is the power of a flare which would be produced by material falling onto a neutron star. Perhaps some of the material from the collapsing core had been ejected into an orbit which had decayed, so that the material had fallen back.

Another possible 'flare' was seen at about the same time as the Ginga flare. JANZOS is the acronym for the Japanese, Australian and New Zealand Observations of the Supernova. It is a gamma-ray observatory set up at Black Birch, New Zealand at an altitude of 1640 m. It consists of two experiments designed to look for very energetic gamma-rays from SN 1987A.

Although the gamma-rays from celestial objects do not penetrate to ground level (which is why rockets, balloons and satellites are used to detect them), the effects of the most energetic gamma-rays as they collide with the atmosphere can be detected from mountain altitudes.

When a gamma-ray of energy between 1–300 TeV (the gamma-rays in this range have energies of ergs†), strikes the upper atmosphere, it transforms into a

Fig. 73. A gamma-ray from SN 1987A interacts with the Earth's atmosphere and produces an 'air shower' of charged particles. The charged particles produce Cerenkov radiation which can be detected by optical telescopes on mountains; some charged particles reach the ground and can themselves be detected. Diagram supplied by E. Budding (Carter Observatory NZ).

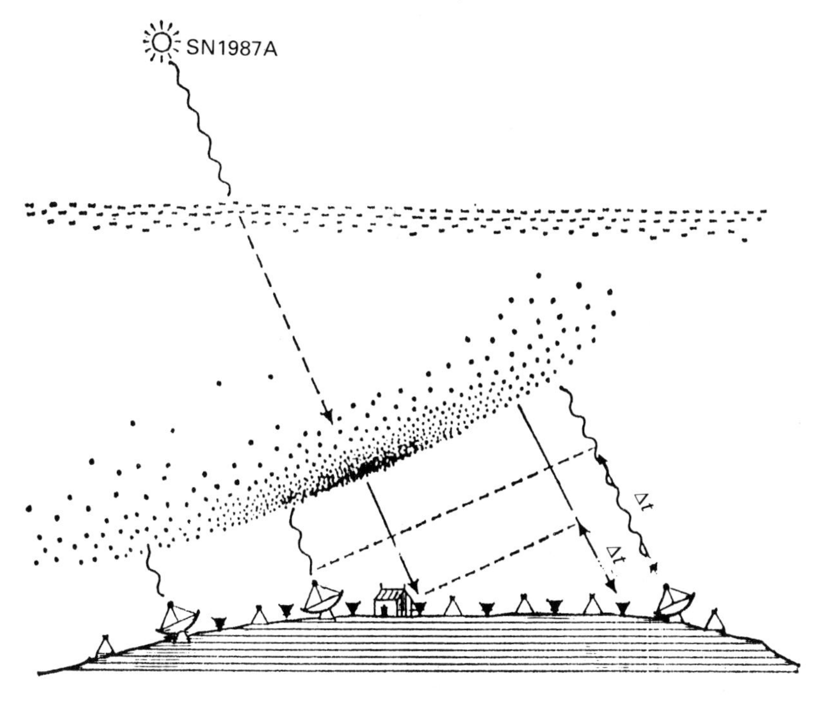

† 1 Tera electron-volt $= 10^{12}$ eV
1 eV $= 1.6 \times 10^{-12}$ erg
$= 1.6 \times 10^{-19}$ J.

swarm of electrons, positrons and photons called an 'air shower'. There is a knock-on effect as the electrons and positrons hit other air molecules, and the shower is amplified to a height of 3 km but is quenched by the greater density of the air as it nears sea level. The charged particles can be detected by scintillation counters on mountains. Moreover, the charged particles create Cerenkov radiation by moving faster than the speed of light through air, and can be detected by optical telescopes. The JANZOS experiment contains both these sorts of detectors, including three 2 m telescopes made of polished aluminium metal mirrors which stand in a frame fixed to view the supernova as it daily crosses the meridian south of the site. In succession ten photomultipliers in a

Fig. 74. The telescopes of the JANZOS experiment are simple structures with a large primary mirror which points in a fixed direction to the south. The image of the supernova tracks across an array of downward-looking photomultiplier tubes positioned at the focus of the mirror. Diagram supplied by E. Budding (Carter Observatory NZ).

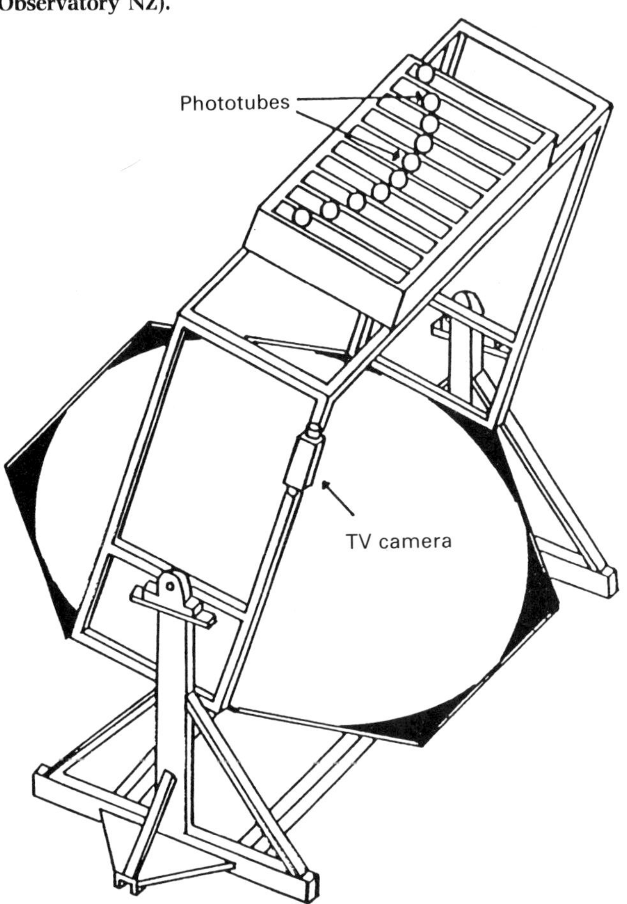

row can see Cerenkov radiation from gamma-rays from the supernova over three hours each night.

The JANZOS telescopes viewed SN 1987A for 14 nights during 1987 December to 1988 January. For two successive nights (1988 January 14 and 15) the telescopes saw what the collaboration (Saktara 1988; Bond *et al.* 1988) interpreted as a weak signal from the direction of the supernova, but nothing outside those two nights. It may be that they glimpsed the supernova's pulsar through a break in the clouds of the supernova envelope which cover it.

The pulsar emerges . . . once

Almost exactly two years after SN 1987A exploded came the first report (IAUC 4735, dated 1989 February 8) of a glimpse of the flashing pulses of a pulsar.

'We report observation of an optical pulsar in the supernova 1987A in the Large Magellanic Cloud,' wrote John Middleditch and his 13 collaborators.

Fig. 75. High energy gamma rays from the supernova may have been detected by JANZOS on two days in 1988 January (two points near Day 325 lie significantly above the level of 1.0 which represents the background level of light seen by the telescopes). The dashed curve represents the X-ray 'light curve' recorded by Ginga at the same time (Figure 55). Diagram supplied by E. Budding (Carter Observatory NZ).

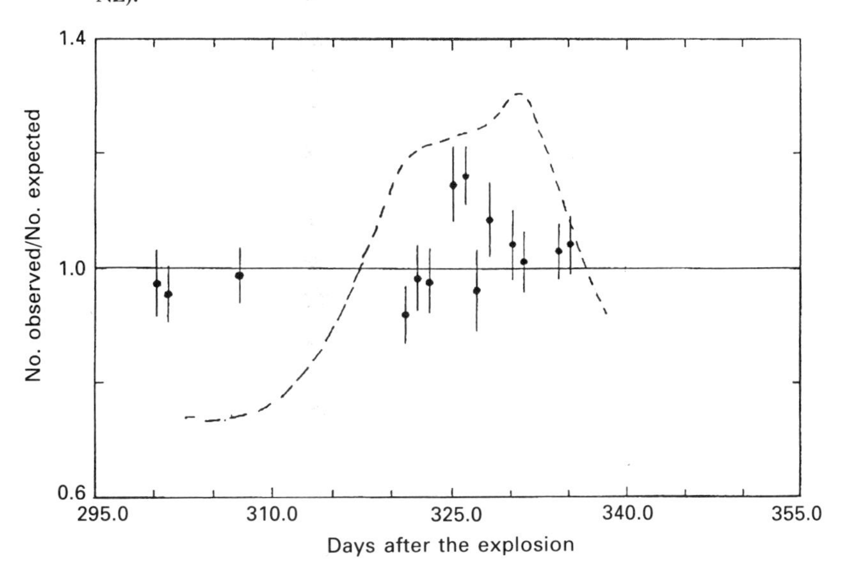

'Detection was made at the Cerro Tololo 4-m telescope on January 18.' The team used a silicon diode detector, responding to light and infra-red radiation of wavelength between 0.4 and 1.0 micrometres. Coaxing the 'somewhat fragile system' into working, they exposed it to the setting supernova with the 4 m telescope in peril as it leant to the horizon. The supernova was then eleventh magnitude. By far most of the signal detected was from the supernova itself, still powered by radioactive decay of nickel and cobalt created in the explosion. But, by analysing the light in a computer for periodic signals (Middleditch 1989), the team discovered that one thousandth of the supernova's light was pulsing periodically – a flickering glow-worm concealed within a bright search-light. Even so the pulsing signal was much brighter than previous limits – Middleditch didn't believe any of it: 'nice if true' he thought. He became convinced that it was, when no trace of the signal appeared in control observations.

The pulsar averaged magnitude 18 to 19 during the seven hour observation (Fig. 76). It pulsed at an astonishing 1968.629 cycles per second, nearly 2000 Hz. It repeated its light pulsations every half millisecond (0.0005 s). This beat by a factor of 2 the previous record for the fastest pulsar and by a factor of 60 the 30 cycles per second of the Crab Pulsar. If we could hear the flashes as

Fig. 76 Observations of the magnitude of the pulsar in SN 1987A show it varying between 19 and 18 at its discovery, but subsequent observations (plotted as upper limits) failed to detect it. Insert: the pulsar light curve (plotted over two cycles) as it flashes twice each cycle. Diagram by R.N. Manchester, B.A. Peterson, A.J. Kalnajs and J.H. Stevenson (private communication).

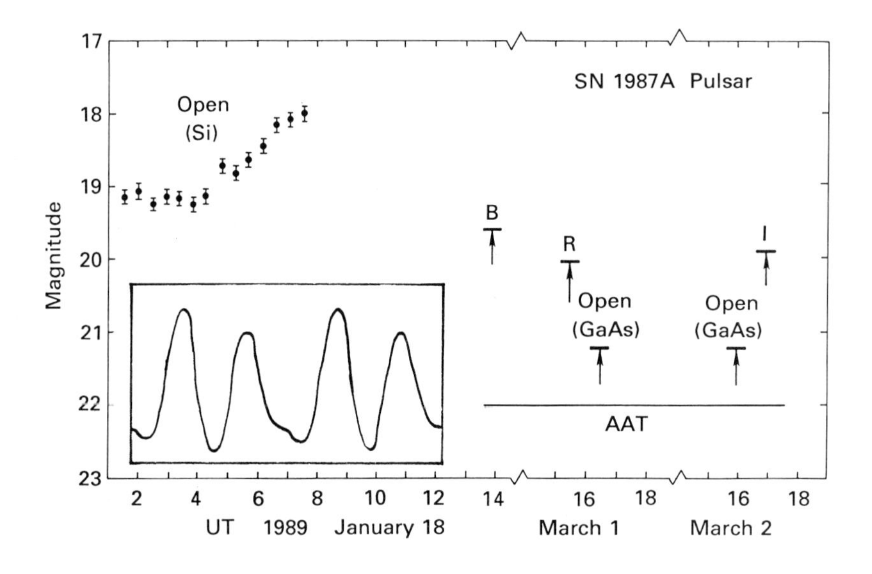

sound, the Crab Pulsar would rumble amongst the bass notes of a piano, 22 keys to the left of middle C; the new pulsar would be a high two octaves above middle C.

The pulsar has, however, only been seen on the one occasion. The pulsar was gradually brightening when recorded by the 4 m CTIO telescope on 18 January. (The California astronomers had made a mess-up of their travel arrangements and arrived at the telescope on the wrong nights. They almost got the time they thought they had, since one of the scheduled observers was hospitalised with seafood poisoning, but in the event had to idle a few days away until their run, including 18th January. It will never be known if the pulsar was 'on' for the nights they thought they had and nearly observed on.)

Observations with the Las Campanas 2.5 m telescope 13 days later revealed nothing. Negative observations with a blue sensitive detector have been reported from Australia, on the same night as the detection. Other negative observations were made on subsequent occasions with both blue and red sensitive detectors both in Australia with the AAT and the Mt Stromlo Observatory Advanced Technology Telescope (ATT) and in Chile at ESO. The negative observations were well below the level of the detection by the Cerro Tololo group (Fig. 76).

Like the JANZOS single detection, this may mean that the pulsar is intermittently visible, the material of the supernova still swirling in clouds across it and hiding it from view, or even falling onto it and quenching the pulsar mechanism. The thought that the Cerro Tololo observation was an instrumental effect has also crossed everyone's mind. In his defence, Middleditch has pointed out not only that reference observations taken the same night on non-pulsing stars showed no similar effects, but also that the pulse shape of the signal, which was modulated with the effect of the Earth's rotation, is very like other pulsars'. Computer reconstruction of the pulsar's pulse shape from the power spectrum yielded a double-peaked light curve, with two flashes separated by 0.2 ms every 0.5 ms. The Crab and Vela Pulsars have similar double flashes every cycle, like light-houses which flash a repeated code of multiple flashes for identification purposes.

Another difficulty has been the frequency of the pulsar – it was on the limit of what theoreticians have expected to be possible.

The speed of the equator of the neutron star in SN 1987A would be very fast if the pulsar flashes at 2000 cycles per second. If its radius is 10 km, its circumference is 60 km and its equatorial speed is 60 km/0.0005 s, that is 0.4 of the speed of light! Obviously its radius cannot be more than about 25 km, or else its equator will move faster than the speed of light.

It is possible for the pulsar's size to be this small. Astronomers have two competing theories for the properties of neutron star material. If the material is 'soft', neutron stars are smaller than if the material is 'stiff' and less easily compressible. The speed of light restriction from the new, fast pulsar implied that it was small and would be broadly compatible with the soft theory. There would still be the difficulty that centrifugal force at the equator of the pulsar would tend to break the pulsar up, just as the rotation speed of Jupiter flattens it at its poles. A half millisecond period of a pulsar is right at the limit of calculated possibilities.

Another pair of competing theories on which the pulsar could provide a lead is the shape of the light beam from pulsars. Either light beams are pencil-shaped (like a well-focussed torch) or fan-beamed. If pulsars have pencil beams then it is somewhat unlikely (about 5 per cent probability) that a given pulsar can be seen from Earth. But if they have fan beams, then it is much more likely than not that a particular pulsar will be detected. The detection of the pulsar in SN 1987A would favour the fan beam theory (although it would obviously not be absolute proof).

An unexpected feature of the Cerro Tololo observations has been the claim of a periodic oscillation in the frequency of the new pulsar. In radio engineers' language, the 2 kHz carrier was frequency modulated. Although the discovery observations were only seven hours long, they indicated the existence of an eight hour oscillation in the frequency, with an amplitude (peak to peak) of 0.003 Hz.

The first thought that came to mind was that this was a binary star period and that the pulsar was one of two stars in orbit around one another. The modulation is amazingly accurately what would be expected from a circular orbit – 'Though not in the habit of talking to myself, I uttered an audible "wow" when I saw the curve,' wrote Middleditch (1989). Calculations show that this hypothesis threw up more problems than it solved. The mass of the companion would be very small, much less than the pulsar itself which is about 1.4 solar masses, if it is like other neutron stars which astronomers know about. In fact the companion, if there is one, would be a planetary mass rather than a stellar one. Either this, or the orbit of the system is pole-on towards the Earth to an accuracy measured in arc minutes rather than degrees. This is very unlikely.

Quite apart from these calculations, there is the doubt that any binary system as light as this could survive the disruption of one component exploding and ejecting about 10 solar masses of itself, particularly as the two stars in the hypothetical binary system are so close that the companion would be, before the explosion, well inside the exploding star. Maybe some of the supernova fell back from the explosion and went into orbit around the neutron star.

An alternative explanation of the frequency modulation is that the pulsar is wobbling in the same way that a top wobbles in response to the force of gravity, its weight levering around the tip. The pulsar would be similarly precessing, also in response to some externally applied couple, presumably the interaction of its magnetic field with the zone which surrounds the pulsar.

Another surprise was that the pulsar is spinning much faster than expected from the theory of its formation. The calculation on p. 208 predicts that the pulsar's period would be measured in seconds. We can run the calculation backwards from the observed frequency of the pulsar and determine Sk – 69 202's rotation period. If angular momentum is conserved and the core of the collapsing supernova progenitor decreases in size by a factor of 100–1000 (from a few thousand kilometres radius to about 10 km), its rotation period would decrease by a factor between 10 000 and a million. If the pulsar's period is 0.0005 s, the progenitor's core's period was at most 10 min. This would be very fast for a supergiant star like Sk – 69 202, which, astronomers would expect, would have a rotation period in excess of days.

At the same time that the pulsar is claimed to be spinning very fast, the power which it emits is indeed, as inferred from the lack of deviation from an exponential decay of SN 1987A's light curve, very small. Compare the new pulsar with the Crab Pulsar and the Vela Pulsar. The Vela Pulsar was created some 10 000 years ago and it now rotates 11 times per second, just three times slower than the Crab Pulsar, aged 1000 years. But the Vela Pulsar's optical magnitude is 10 000 times fainter than the Crab Pulsar's (allowing for its different distance). If the same law is applied to the new pulsar in SN 1987A it would be millions of times brighter than it is – a naked eye object, outshining the rest of the supernova at the time of the pulsar's discovery.

Since the power radiated by a neutron star depends on its speed of rotation and its magnetic field, and the new pulsar's speed of rotation was claimed to be very fast, its magnetic field would be weak.

The status of SN 1987A's neutron star remained, at the time of the proof stage of this book (July 1989), a puzzle.

13 A beacon across space

The supernova as an intergalactic probe

The region between the stars is interstellar space; it contains galactic material which has not made stars yet and material which has been ejected from stars, having been enriched with elements which they have manufactured.

The region between galaxies is intergalactic space. It may contain material which has never been in any galaxy and which has been left over from the Big Bang (pristine material). It may also contain material ejected by galaxies: material torn by intergalactic tides from one galaxy or another; material ejected by the supernovae in a galaxy; material which has floated free from the outer halo of a galaxy. About such material in a galactic wind little is known.

The method used to probe the intergalactic regions is to view a star in a galaxy outside our own and to look at the effect on the star's spectrum of intergalactic material. The star must be comparatively bright to make it worthwhile studying the expectedly small effects on its spectrum. Quasars can be used to probe intergalactic space, and so too can supernovae in other galaxies.

The LMC supernova is the brightest star studied over an intergalactic path length. Its light has probed the interstellar and intergalactic space between the LMC and us and contains the absorption spectrum of the gases which inhabit these regions.

Since the supernova is so bright, comparatively, it has been possible to obtain very accurate spectra of very weak absorption lines from rare elements, or from elements whose spectral features are weak. Normally astronomical spectra are made by exposing starlight for a few hours, exceptionally for a few nights. In the case of a supernova the exposure time is limited by the time that the supernova exists, so there is an extreme upper limit of, say, months of exposure time; in practice, since other astronomers demand their fair share of access to the telescopes to carry out observations for their particular interest this extreme is never met in practice and a more likely upper limit would be 20 hours.

Obviously, the brighter the star the more light can be detected in the time available. For any astronomer interested in probing the intergalactic regions of space, SN 1987A was more than 100 times brighter than any previous supernova in recent time, and spectra more than 10 times more accurate can be obtained – or spectra with 100 times finer resolution, if the same accuracy can be tolerated. This was the fundamental reason why the Peter Gillingham spectrograph was built to respond to the supernova (Chapter 3).

Intergalactic clouds and the Magellanic Stream

The interstellar and intergalactic material does not occupy space uniformly. It clumps in clouds. So the absorptions of the clouds, which are strung on the line of sight to the supernova, like beads on a necklace, show as individual spectral lines.

The clouds themselves, however, are in orbit in, around and between our Galaxy and the LMC. They have different velocities in the line of sight and the Doppler shift moves their absorptions to slightly different wavelengths.

The clouds which are nearby in our Galaxy are moving circularly around the Galaxy and so too is our Sun. Thus there is little relative motion between nearby interstellar clouds and the Sun. The radial velocity of the clouds observed from Earth shows predominantly the effect of the motion of the Earth around the Sun. Once the terrestrial motion is corrected for, the nearby clouds have zero radial velocity relative to us.

Those interstellar clouds in our Galaxy, which orbit in more distant parts of its plane, would have larger motions relative to the Sun, but the line of sight from us to the LMC supernova does not pass through such regions. The LMC lies in a direction which is out of the galactic plane.

Any clouds in the LMC itself show its recession velocity (270 km/s), with a slight variation due to the rotation of regions of the LMC around its centre of mass. Only the parts of the LMC penetrated by the line of sight contribute to the velocities seen and they fall in a narrow range, given by the rotational velocity of the LMC in that direction.

Any clouds which are moving between the LMC and our Galaxy show up with an intermediate radial velocity.

Each cloud in the line of sight thus shows as a narrow spectral line in a range of velocities between 0 km/s and the 300 km/s radial velocity of the LMC.

Amongst the elements which have been detected in this way are sodium and calcium. Vidal-Madjar *et al.* (1987) have identified 40 individual clouds of interstellar calcium between us and the supernova. About six are in our own Galaxy, a dozen are in the LMC and the remainder lie in a tidal zone between, bridging the gap.

It is a little more difficult to identify individual clouds of interstellar sodium. The sodium D line is actually a pair of lines, D1 and D2, separated by 2 Å; the range of 300 km/s for the radial velocity of the clouds is 5Å. Thus the band of absorptions from the D1 spectral line overlap and muddle with the band of absorptions of the D2 line. But by careful analysis it is possible to match the clouds of sodium as measured in the two spectral lines.

Matching sodium with calcium, however, produces a different story. There are sodium clouds at low velocities, corresponding to the calcium clouds in our

Fig. 77. Spectra of interstellar potassium (top), calcium (middle) and sodium (bottom), superimposed on a common scale of radial velocity relative to the Sun (bottom horizontal axis), and with 24 interstellar clouds marked by ticks (top horizontal axis). The shapes of the sodium and calcium spectra are very similar, except that the sodium clouds between 50 and 250 km/s are relatively much weaker than the calcium clouds, the phenomenon of 'sodium depletion'. These impressive data are from ESO's splendid Coude Echelle Spectrograph (CES) in its 3.6 m telescope building, fed by the Coudé Auxiliary Telescope (CAT) (Vidal-Madjar *et al.* 1987).

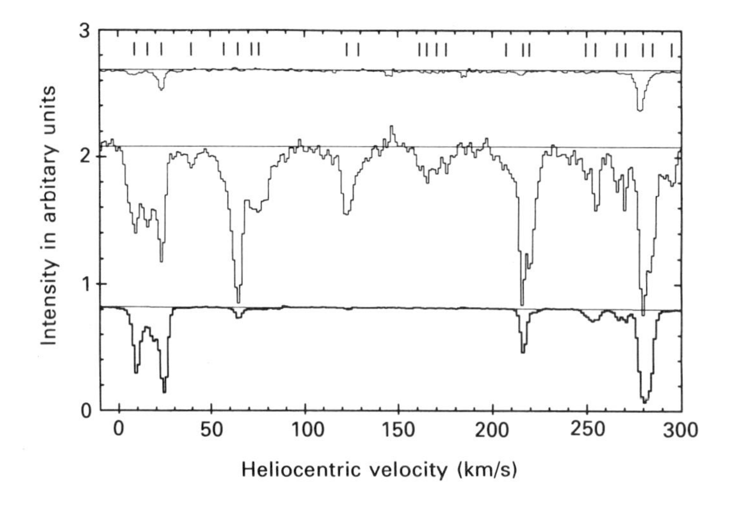

Galaxy. There are higher velocity sodium clouds which correspond well with the calcium clouds in the LMC. But the intermediate velocity sodium clouds, corresponding to the intergalactic calcium clouds, are missing.

At first sight, this means that intergalactic clouds are of a different composition from the ones in either of the two galaxies. Intergalactic clouds seemed to have a deficiency of sodium.

There is a more subtle reason for this discovery. It is that Galactic and Magellanic interstellar clouds have a deficiency of calcium. This deficiency is only apparent. Galactic and Magellanic clouds are not really deficient in calcium: the calcium is hiding.

The interstellar clouds in our Galaxy and in the LMC contain dust which has been ejected into interstellar space by old red supergiant stars. Calcium atoms stick to the dust grains. The calcium atoms are then inhibited from absorbing light and do not show as absorption lines. Thus, in our very dusty Galaxy, and even in the less dusty LMC, calcium is depleted from the interstellar gas and deposited on the interstellar grains.

Calcium should show much more strongly than sodium, because its spectral lines are much more readily produced than sodium's and they are equally abundant. Calcium does not show more strongly – calcium and sodium are both equally visible because of the calcium depletion.

In intergalactic space, however, the clouds evidently have less dust. Calcium is not depleted. The readiness of the calcium atoms to create an absorption line compensates for the fact that there are intrinsically less atoms in intergalactic space. But sodium does not have this compensation factor operating. Therefore, in the intergalactic clouds, calcium shows more strongly than sodium.

The origin of these intergalactic clouds is a matter for debate.

There is a large number of clouds, strung out uniformly between our Galaxy and the LMC. Did they get ejected from our Galaxy or the LMC or both? Were they accreted to our Galaxy like a shower of meteorites?

The ratio of calcium to sodium varies in a vary regular way from cloud to cloud. This means that there is a smooth change in physical properties like depletion from one cloud to the next. This, in turn, points to a large scale event for their origin, rather than numerous random processes. This favours a tidal theory, with clouds pulled from one galaxy or the other.

This idea is reminiscent of the 'Magellanic Stream'. This is a thread of neutral hydrogen clouds which orbit our Galaxy and both the LMC and the SMC. It is thought that the Magellanic Stream was formed by the tidal interaction between our Galaxy and the Magellanic Clouds in orbit around it.

The temperatures of space – hot and cold

The gas clouds are rather cold, with temperatures less than 170 K according to observations of the sodium D2 line made by Max Pettini and his colleagues with the Peter Gillingham spectrograph.

The sodium D line is doubled into the pair called D1 and D2, but each of these is doubled again by a phenomenon called hyperfine structure. This is caused by an interaction of the sodium nucleus with its orbitting electrons. The sodium nucleus has a spin of $\frac{3}{2}$ and its atomic electrons have a spin $\frac{1}{2}$. The electrons like to lie with their spin parallel or anti-parallel to the nucleus, and this causes a slight difference in their energy levels. Those electrons which make the transition from one of these states absorb slightly different amounts of energy from those making a transition from the other state. The spectral lines are split by the hyperfine structure. By observing with a spectrograph with enough resolution, the hyperfine structure can be distinguished.

But hyperfine structure can only be distinguished if the sodium atoms are not moving much. If they are, the Doppler shift of the individual atoms blurs the hyperfine structure and it cannot be resolved.

What reasons are there for the sodium atoms to move? They could be in distinct blobs of sodium which are stirred up in a turbulent fashion. Or they could be in a hot gas so that thermal motion causes the individual atoms to rush about within a cloud.

Using the newly made AAT spectograph, Pettini (1988) obtained data in which the hyperfine structure can be seen. In fact the atomic hyperfine structure in the sodium D2 line shows that velocities in the gas clouds are less than 0.8 km/s. If this is due to the lack of thermal motion in the sodium of the intergalactic clouds then the temperature within the clouds is less than 170 K.

At the same time, Pettini has also shown that the cold clouds live in a hot bath.

Pettini *et al.* (1988) have detected an unusual absorption line in the supernova which has its origin in the ion of iron which is nine times ionised, Fe^{9+}. (In astronomers' notation this is [FeX].) It takes a high temperature to knock the ninth electron off an iron atom, so this spectral line is produced in gas at a temperature of 2 million degrees Kelvin. As an emission line it is detected in the hot corona of the Sun. This indicates that the temperature of the solar atmosphere is about 2 million degrees Kelvin. This is why the Sun is a strong source of X-rays, which are emitted by hot gases at this kind of temperature.

Pettini has discovered the spectral line of FeX in absorption in the spectrum of the supernova. Its velocity indicates that it is predominantly in the LMC but there is a possible component at the zero velocity of hot gas in our Galaxy. Naturally the spectral line is broad because the thermal motion of the iron ions is very large.

Some of the hot gas is clearly within the LMC and could have been the result of earlier supernovae. Apparently the number in this region of the LMC is much higher than average (something like 15 per century) – this must be associated with the star formation activity connected with 30 Doradus and the nebulae near SN 1987A. But some of the hot gas originates in the halo around our own

Fig. 78. The spectrum of the sodium D2 line towards the supernova shows several absorption components from individual gas clouds along the line of sight, here identified by their radial velocities (1–65 km/s). The less intense, unsaturated lines all show the same splitting into two unequal components (0.022 Å apart) which are the hyperfine structure of the D2 transition associated with the degree of freedom for the spin angular momentum of the sodium nucleus. The $^2S_{\frac{1}{2}}$ level in the atom is split by interaction with the spin $I=\frac{3}{2}$ of the nucleus, into components in the ratio $I:I+1$, i.e. 3 : 5. The fact that the hyperfine structure is resolved implies that turbulent motion within individual clouds is less than 0.8 km/s and that the thermal motion of the sodium atoms is therefore less than 170 K. Data from Max Pettini, using the improvised spectrograph of the AAT.

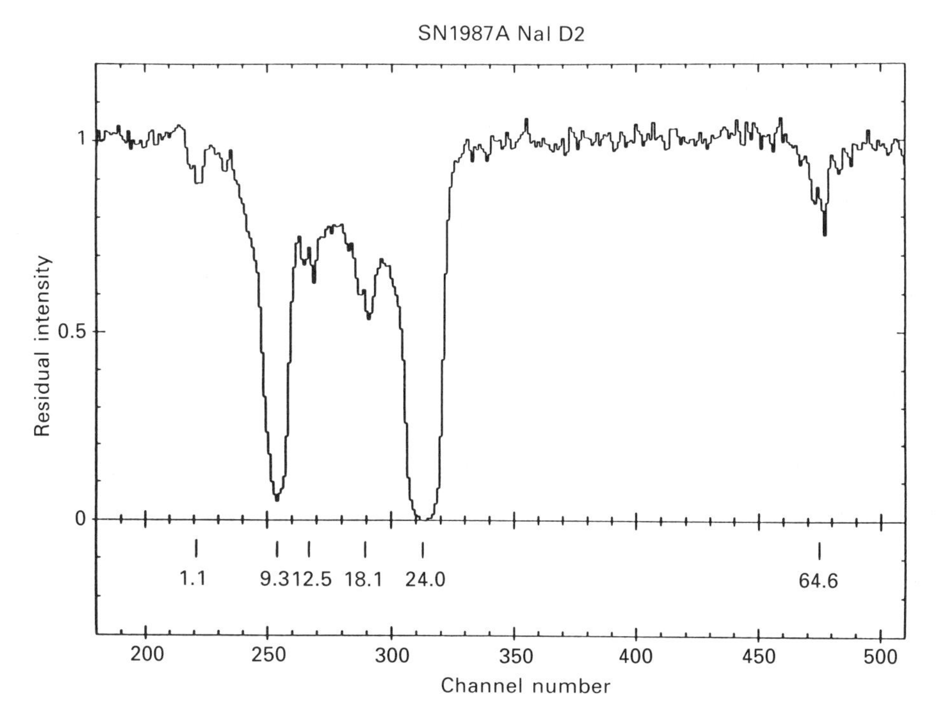

Galaxy. If it is uniformly distributed around our Galaxy, then the hot halo contains as much mass as in the stars of the Galaxy. How the sodium clouds remain cold in such a hot gas is as much of a mystery as where the hot gas comes from.

Pristine material

Lithium is an important element for the theories of element creation. Lithium was formed in the Big Bang and is still being formed in some reactions on the surface of supernovae; but its two isotopes are destroyed at the temperatures over 10 million degrees Kelvin which are found inside stars. Interstellar material has been cycled through stars as they form from the interstellar medium. The material is heated to temperatures well in excess of 10 million degrees Kelvin inside the stars. In the later stages of their lives, the stars eject material back into space.

Thus the abundance of lithium in the interstellar material has been altered from its cosmic value. The abundance of lithium varies dramatically from place

Fig. 79. A busy part of the red spectrum of the supernova is confused by absorption lines arising in the Earth's atmosphere (marked by a crossed circle symbol) and Diffuse Interstellar Bands (DIBs); but the most prominent feature marked LMC [FeX] is due to nine-times ionised iron in the LMC. AAT data by Max Pettini and his collaborators.

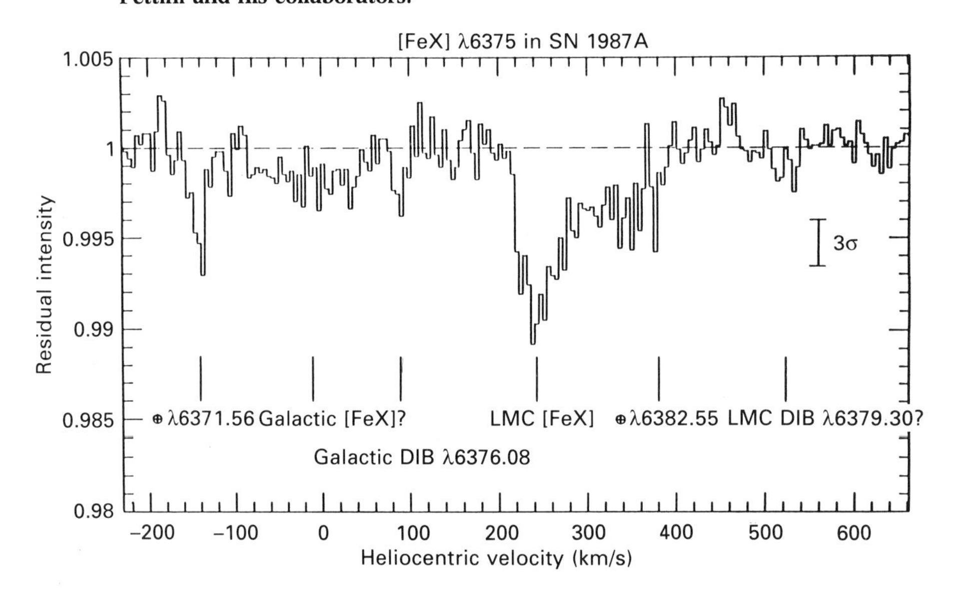

to place in our Galaxy – it depends on the history of each area. Astronomers hoped that if they observed places with as little history as possible then they could see primordial lithium. With this in mind they have tried to identify lithium on the surfaces of inactive, unevolved old stars, like those in globular clusters. Similarly, astronomers would expect that intergalactic clouds had, equally, not been re-cycled much.

It has not been possible to detect lithium in the direction towards the supernova. Thus the intergalactic clouds have a lithium abundance of less than the old-star value (Baade and Magain 1988). Old stars are evidently not so pristine as they were thought to be.

Other elements were detected towards SN 1987A by IUE. Among them were carbon, oxygen, magnesium, aluminium, silicon, sulphur, chlorine, chromium, manganese, iron, nickel and zinc, some of them detected in the interstellar medium for the first time (Blades *et al.* 1988).

14
SN 1987A and our next supernova

The nearest supernovae

There are probably at least 10^8 neutron stars in our Galaxy, of which 10^5 are radio pulsars (only 10^3 have been detected) and about 20 are X-ray pulsars (Burrows 1988). About 1000 of these neutron stars are within 100 pc (300 light-years) of the Earth. If they were all formed by Type II supernovae, then there have been 500 Type II supernovae of magnitude -13 or brighter in the lifetime of the Earth (which is half as old as the Galaxy). This is the equivalent to one every 50 million years.

The brightest supernovae of all occur within a distance of about 10 pc (30 light-years), 5000 times closer than the LMC. There may have been a few such cases in the lifetime of the Sun (5×10^9 years).

The Sun is (by definition) of mass 1 solar mass; it will therefore not become a Type II supernova (which are from stars more massive than about 7 solar masses). Nor will it become a Type I supernova since it is not a white dwarf in a binary star system. There is no likelihood of us experiencing a supernova closer than about 10 pc.

If we did experience a supernova at 10 pc, we would be able to detect it first by the flood of neutrinos, 25 million times more than from SN 1987A. One hundred million neutrinos would be detected in each of the IMB and Kamioka detectors: they certainly would not be able to count such a number of events, which would produce a burst of light of 10^{10} photons in each 5000 tonne tank.

One hour or so afterwards there would be a very bright brief flash of ultraviolet radiation from the supernova, as the shock broke out from the star's surface. If the night-time side of the Earth faced the supernova, light associated with the ultraviolet flash would certainly be noticed by a casual observer.

Ionisation of the atmosphere would cause a temporary breakdown of radio communications.

Over the next day the supernova would brighten to outshine the Moon and rival the Sun. It would remain visible to the unaided eye for some years.

The supernova would grow in size at the rate of 1° in six years. Thus it would be visible as a perceptible disc within a year and the size of the Moon in three years.

The pulsar in its interior would become visible to the naked eye; but, unless it flashed slower than a few times per second, it would pulse too rapidly to be seen to flash by eye.

In 300 years the supernova would have expanded enough for the outer parts to have encompassed the Earth. Cosmic radiation would increase, aurorae would be more frequent. Species would mutate more rapidly due to an increased frequency of radiation-induced genetic abnormality but there would be no significant risk of radioactivity for individuals.

The fate of a supernova's planetary system

There is no evidence that Sk − 69 202 had a planetary system. If it had one, then its inner planets would have been encompassed and evaporated during the increase in size of the star while it was in its red supergiant stage. Its outer planets' ecology would have been devastated by the change in luminosity and temperature during its evolution from blue Main Sequence star to red supergiant and then to blue supergiant.

It seems unlikely that any creatures on these planets had managed to evolve during the brief stable life of the star (10^6 years is 1000 times shorter than the time it took life to begin on Earth). If they did, and survived the unstable advanced stages in Sk − 69 202's life, then the creatures witnessed the supernova explosion from close at hand.

Suppose the extraterrestrial creatures were water-based life forms like us, each with a body of 100 kg. The flood of neutrinos from a supernova at the distance of a sun would induce 10^{16} interactions with electrons or protons in each body. Each of these interactions would generate a beta-ray, and in total about 10^9 erg of energy from ionising radiation would be deposited in the creature's body. This energy per 100 kg is equivalent to a dose of radiation of 10 000 rads (1 rad is equivalent to 1 roentgen, for practical purposes, and the medical effects of radiation are calibrated in roentgens). In a lifetime human beings are exposed to about 5 roentgens of natural radioactivity from cosmic rays, terrestrial rocks etc. and suffer no particularly harmful effects. An exposure of 100 roentgens from a nuclear weapon causes people to suffer from radiation sickness, but most recover. At a dose of 1000 roentgens, however, most people eventually die in a matter of months. At the dose of 10 000 roentgens just calculated as typical of the radiation induced by neutrinos on creatures on a nearby planet, human beings are incapacitated immediately and die within a week.

If the extraterrestrial creatures survived the neutrino burst, but lived on the daylight side of their planet at the moment of the ultraviolet flash of the shock breakout, then they were burnt to death by the extreme sunburn. The heating of the atmosphere on the day-time side of the planet created hurricane-force winds on the night-time side, with consequential storm devastation.

After the planet had turned the night-time side towards the exploding star a few hours later, the residual survivors could see their sun enlarging in size, growing to the appearance of an on-rushing wall of fire by the end of the day. It engulfed them, probably during the next day or two. They and their planet were vaporised.

The end of such a planet and its lifeforms seems assured and relatively quick by one method or another. In any case there is no time to worry about one's star becoming a supernova: it happens without warning.

Why is there no warning of a supernova?

The energy from the centre of any star diffuses outwards very slowly. In the interior of the star, the gamma-rays and X-rays encounter dense material and move many times back and forth and only gradually make overall progress outwards. The energy created in the core thus stays within the star for millennia. Eventually, however, the energy drifts upwards to the surface of the star and is emitted, as ultraviolet, light and infrared. On the other hand, the last stages of the life of a massive star last a much shorter time than millennia: silicon burning persists only a day or so before core collapse. At the moment of core collapse, therefore, the heat from silicon burning has not diffused very far from the core. Any effects which it might have had on the surface emissions never take place, before the much more dramatic effects of the core collapse totally dominate the situation! Because the effects of silicon burning never reach the surface of the star no 'messages' are signalled by the surface of the star relating to the silicon burning within.

There is thus no immediate warning of a supernova.

We can estimate which stars in our sky are candidates to become supernovae, but not which ones are imminent.

Table 28. *Supernovae in nearby galaxies*[a]

Galaxy	Distance	Absorption (mag)	Supernova brightness	Mean time between SN (years)
Our Galaxy	10 pc	0	−18	2×10^9
	100 pc	0	−13	2×10^7
	1000pc	1	−7	1000
	5000 pc	5	−0.5	200
	10000 pc	10	+7	100
	anywhere			50
LMC	50 kpc	1	+1	500
SMC	55 kpc	1	+2	4000
M 31	600 kpc	1?	+7	30

[a]Typical values assumed for a supernova which reaches absolute magnitude −18. One kiloparsec (kpc) is just over 3000 light-years. Galaxy distances are approximate.

All the massive stars in the sky will be Type II supernovae one day. This includes, for example all the bright stars in the constellation Orion. Betelgeuse is the star of this constellation which is the most advanced in its evolutionary path and, a red supergiant, is the kind of star believed most typical of a supernova progenitor.

The next naked eye supernova

If there is no warning of a supernova and astronomers cannot predict which star will next become a supernova, can we estimate how long we must wait for the next?

It is 380 years since the last naked eye supernova in 1604, and over 300 years since the supernova which produced the supernova remnant Cas A, the last supernova known in our Galaxy. Will we have to wait centuries for the next bright supernova?

The answer is that the next naked eye supernova could be tomorrow. When Halley's Comet visited the vicinity of the Earth recently, adults who saw it knew that there would not be a second chance for them to see it in their lifetime. The comet is periodic and its next return can be predicted in advance, so well that the very day and hour of its closest approach to the Sun can be estimated.

But supernovae are not periodic and forecastable like comets. One supernova is not correlated with the next. All that is established is the approximate value of the average number per century in our Galaxy and the nearby ones, and the

Table 29. *Bright supernovae in historic times*

Date	Radio remnant	Brightness
185 AD	G315.4–2.3	−6
386	G11.2–0.3	+2?
393	?	?
1006	G327.6+14.5	−9
1054	Crab Nebula	−5
1181	3C58	+1
1572	Tycho's SNR	−4.0
1604	Kepler's SNR	−2.6
1667	Cas A	+6?
1987	not yet known	+2.8

actual interval between one and the next is a statistical variable which, in principle may have any value from zero to infinity, with a certain probability for each value between. A one day interval between one naked eye supernova and the next may be unlikely but it is not impossible.

The above words are hopeful and optimistic; however, the realistic chances of seeing a naked eye supernova in a lifetime are not good.

Representative values of the calculated mean time between supernovae of a given brightness is given in Table 28. They can be checked for the brighter end by comparing with Table 29. There have been four galactic supernovae with magnitudes brighter than zero observed in 1000 years, in good agreement with the calculations. The brightest supernovae can appear only in our Galaxy, and the chances of seeing a supernova in the LMC are poor; the chances of seeing one in the SMC poorer still. To be naked eye brightness, a supernova in the Andromeda Galaxy would have to be unusually bright and occur in a region virtually free of interstellar absorption: even then, it would only just make it above binocular brightness.

Realistically, those people who saw with their own eyes the supernova in the LMC are likely not to repeat their experience. For scientists its study is a once-in-a-career opportunity that puts them in the distinguished company of Tycho Brahe, Johannes Kepler and Galileo.

Appendix 1 Calendar for the supernova

Add the day of the month to the table entry in order to determine the time elapsed since 1987 February 23; 1987 February 23 at 07:35:42 UT corresponds to Julian Day 2446849.816458.

	1987	1988	1989	1990	1991
January 0		311	677	1042	1407
February 0	−23	342	708	1073	1438
March 0	5	371	736	1101	1466
April 0	36	402	767	1132	1497
May 0	66	432	797	1162	1527
June 0	97	463	828	1193	1558
July 0	127	493	858	1223	1588
August 0	158	524	889	1254	1619
September 0	189	555	920	1285	1650
October 0	219	585	950	1315	1680
November 0	250	616	981	1346	1711
December 0	280	646	1011	1376	1741

Bibliography

References

This book is not intended to provide my fellow astronomers with an exhaustive review of the vast literature about the supernova, but to make sense in an orderly manner of its story. However, I have felt that it may be useful if I provide some entry points into the scientific literature and if I support statements about the history of the discoveries by references, particularly to the more unusual journals. The two conference proceedings referenced here as the *ESO Workshop* and the *George Mason Workshop* were particularly influential scientific publications produced by their contributors, editors and publishers with remarkable speed within the first year of the supernova. I have used them as convenient compilations from which to tell my tale.

Aglietta, M. *et al* (1987a). *Europhys Lett.* **3** (12) 1315; also *ESO Workshop*, 207; also *Moriond-22*, 717; also *George Mason Workshop*, 119.

Aglietta, M. *et al.* (1987b). *Europhys Lett.* **3** (12), 1321; also *ESO Workshop*, 207.

Aitken, D. (1988). AAO Symposium 1988 Sep. 28.

Alexayev, E.N., Alexayeva, L.N., Krivosheina, I.V. and Volchenko, V.I. (1987). *ESO Workshop*, 237; also *Moriond-22*, 739.

Allen, D.A. (1987). *AAO Newsletter* No. 41, 2.

Allen, D.A. (1988). *AAO Newsletter* No. 45, 3.

Amaldi, E. *et al.* (1988). *George Mason Workshop*, 453.

Andersen, P.H. (1988). *Physics Today* **41**, 23.

Arnett, W.A. (1988). IAU General Assembly, Baltimore, 1988 August.

Baade, D. and Magain, P. (1988). *Astron Astrophys.* **194**, 237.

Bahcall, J.N., Cleveland, B.T., Davis, R. and Rowley, J.K. (1985). *Ap. J.*, **292**, L79.

Barbiellini, G. and Cocconi, G. (1987). *Nature*, **329**, 125.

Bartell, N. *et al.* (1988). *George Mason Workshop*, 81.

Bates, R. (1988). *George Mason Workshop*, 472.
Beier, E.W. (1986). *Proceedings of Seventh Workshop on GUT*, Toyama.
Bildsten, L. and Wang, J.C.L. (1988). *George Mason Workshop*, 116.
Bionta, R.M. *et al.* (1987). *Phys. Rev. Lett.* **58**, 1494.
Blades, J.C. *et al.* (1988). *George Mason Workshop*, 261.
Blanco, V. (1987a). *IAUC*, 4349.
Blanco, V. (1987b). *ESO Workshop*, 27.
Bond, I.A. *et al.* (1988). *Phys. Rev. Lett.* **60**, 1110.
Boris, S. *et al.* (1985). *Phys. Lett.*, **159B**, 217.
Branch, D. (1987). *ESO Workshop*, 407.
Brown, S.G. (1987). *Sky and Telescope*, **74**, 453.
Burrows, A. (1987a). *ESO Workshop*, 315.
Burrows, A. (1987b). *Physics Today*, **40**, 28.
Burrows, A. (1988). *George Mason Workshop*, 161.
Burrows, A. and Lattimer, J. (1987). *Ap. J.*, **318**, L63.
Cannon, R. (1987). *Sky at Night*, BBC TV, 27 Sept. 1987.
Catchpole, R. *et al.* (1987). *MNRAS*, **229**, 15P.
Cassatella, A. (1987). *ESO Workshop*, 101.
Cassatella, A., Wamsteker, W., Sanz, L. and Gry, C. (1987). *IAUC*, 4330.
Castagnoli, C. (1987). *IAUC*, 4323, 4332.
Chalabaev, A.A., Perrier, C. and Mariotti, J.M. (1988). *George Mason Workshop*, 236.
Chevalier, R. (1987). *ESO Workshop*, 481.
Chevalier, R. and Fransson, C. (1987). *Nature*, **328**, 44.
Chiu, H.Y. (1988). *George Mason Workshop*, 185.
Chu, Y.-H. (1987). *IAUC*, 4322.
Close, F., Marten, M. and Sutton, C. (1987). *The Particle Explosion*, Oxford University Press.
Cook, W.R., Palmer, D., Prince, T., Schindler, S., Starr, C. and Stone, F. (1988). *IAUC*, 4527.
Couch, W. (1988). *George Mason Workshop*, 60.
Cropper, M., Bailey, J., McCowage, J., Cannon, R.D., Couch, W.J., Walsh, J.R., Straede, J.Q. and Freeman, F. (1987). *MNRAS*, **231**, 695.
Crotts, A. (1988). *IAUC*, 4561.
Danziger, I.J. (ed.) (1987). *Proceedings of the European Southern Observatory Workshop on SN1987A*, ESO Conference Proceeding No. 26, Garching. Referred to in this bibliography as *ESO Workshop*.
Danziger, I.J. *et al.* (1988). *George Mason Workshop*, 37.
Dar, A. (1988). *George Mason Workshop*, 220.
Disney, M. (1984). *The Hidden Universe*, J.M. Dent.
Dotani, T. *et al.* (1987). *Nature*,**330**, 230.
Evans, R. (1988). *Southern Astronomy* pilot issue, 25.
Fazio, G.G., Jelley, J.V., Charman, W.N. (1970). *Nature*, **228**, 260.
Feast, M.W. (1988). *George Mason Workshop*, 51.
Fesen, R. (1985). *Astrophys. J.* **297**, L29. See also *Sky and Telescope*, **71**, 550 (1986).
Filippenko, A. (1988). *George Mason Workshop*, 106.
Fosbury, R. (1988). Private communication.
Franklin, J. (1988). *George Mason Workshop*, 197.
Fritschi *et al.* (1986). *Phys. Lett.* **173B**, 485.
Galeotti, P. (1987). *ESO Workshop*, 217.
Gillingham, P. (1987). *Instrumentation for ground-based optical astronomy: present and future* ed. L.B. Robinson p. 134, Springer, NY.

Gilmozzi, R. (1987). *ESO Workshop* 19.
Gilmozzi, R., Cassatella, A., Clavel, J., Fransson, C., Gonzalez, R., Gry, C., Panagia, N., Tallavera, A. and Wamsteker, W. (1987). *Nature*, **328**, 318.
Gonzalez, R. *et al.* (1987). *ESO Workshop*, 33.
Gry, C., Cassatella, A., Wamsteker, W., Sanz, L. and Panagia, N. (1987). *IAUC*, 4327.
Guiffres, C. *et al.* (1988). *ESO preprint*, 591.
Hamuy, M., Suntzeff, N.B., Gonzalez, R. and Martin, G. (1988). Preprint *Supernova 1987A: UBVRI photometry at Cerro Tololo.*
Hanuschik, R.W., Thimm, G. and Dachs, J. (1988). *ESO Messenger*, No. 51, 7.
Hauschildt, P., Spies, W., Wehrse, R. and Shaviv, G. (1987). *ESO Workshop*, 433.
Hazen, M.L. (1987). *IAUC*, 4367.
Helfand, D. (1987). *Physics Today*, **40**, 26.
Hillebrandt, W., Hoflich, P., Truran, J.W. and Weiss, A. (1987). *Nature*, **327**, 597.
Hirata, K. *et al.* (1987). *Phys. Rev. Lett.* **59**, 1490; also *Moriond-22*, 725.
Huchra, J. and Olowin, R. (1988). *IAUC*, 4567.
Humphreys, R.M., Jones, T.J., Davidson, K., Ghigo, F. and Zumach, W. (1987). *IAUC*, 4325.
Itoh, M. *et al.* (1987). *Nature*, **330**, 233.
Jedrzejewski, R. (1987). *Sky and Telescope*, **73**, 470.
Kefatos, M. and Michalitsianos, A. (eds) (1988). *Supernova 1987A in the Large Magellanic Cloud: Fourth George Mason University Astrophysics Workshop, Fairfax, VA 12–14 Oct. 1987*, Cambridge University Press. Referred to in this bibliography as *George Mason Workshop.*
Kajita, T. (1987). *Moriond-22*, 689.
Karovska, M., Nisenson, P., Noyes, R. and Papaliolios, C. (1987). *IAUC*, 4382.
Kirshner, R. (1988a). *Science Now*, BBC Radio-4, 5 Mar. 1988.
Kirshner, R. (1988b). *George Mason Workshop*, 87.
Kirshner, R. *et al.* (1987). *IAUC*, 4435.
Kormendy, J. and Knapp, G.R. (eds) (1987). *Dark Matter in the Universe*, Reidel, Dordrecht.
Koshiba, M. (1987). *ESO Workshop*, 219; also *George Mason Workshop*, 131.
Krauss, L.M. (1987). *Nature*, **329**, 689.
Kuo, T.K. and Pantaleone, J.T. (1988). *George Mason Workshop*, 200.
Larson, H.P. *et al.* (1987). *ESO Workshop*, 74.
Larson, H.P. *et al.* (1988). *George Mason Workshop*, 74.
Lasker, B. (1987). *IAUC*, 4318.
Lasker, B. *et al.* (1988). *George Mason Workshop*, 449.
Lucy, L. (1987). *ESO Workshop*, 417; also *George Mason Workshop*, 323; also *Astr. Astrophys.*, June 1987.
Maeder, A. (1987). *ESO Workshop*, 251.
Mahoney, W.A. *et al.* (1988). *IAUC*, 4584.
Makino, F. (1987a). *IAUC*, 4447.
Makino, F. (1987b). *IAUC*, 4466.
Manchester, R.N. (1987). *ESO Workshop*, 177.
Maran, S.P. (1988). *Smithsonian*, **19**, 46.
Masai, K., Hayakawa, S., Itoh, H. and Nomoto, K. (1987). *Nature*, **330**, 235.
Matz, S.M. *et al.* (1987). *IAUC*, 4510.
Matz, S.M. *et al.* (1988). *Nature*, **331**, 416.
McNaught, R.H. (1987a). *AAO Newsletter*, No. 41.
McNaught, R.H. (1987b). *IAUC*, 4316, 4317.
Meikle, W.P.S. (1988). *AAO Symposium*, 1988, Sep. 28.

Meikle, W.P.S., Matcher, S.J. and Morgan, B.L. (1987). *Nature*, **329**, 608; also *ESO Workshop*, 197.
Menzies, J.W. (1987). *ESO Workshop*, 73.
Menzies, J.W. *et al.* (1988). *MNRAS*, **227**, 49P.
Middleditch, J. (1989) *Computers in Physics*, **3**, 14.
Middleditch, J. *et al.* (1989) *IAUC*, 4735.
Murdin, P. (1987). *Nature*, **329**, 12.
Nagasawa, M., Nakamura, T. and Miyama, S. (1988). *Publ. Astr. Soc. Japan*, in press.
Nisenson, P. *et al.* (1987). *Astrophys. J.*, **320**, L15.
Ögelman, H., Böhringer, H., Buchert, S., Cakir, S., LaBelle, J., Treumann, R.A. (1987). *Astr. Astrophys. Letters*, **183**, L27.
Panagia, N. (1988). IAU General Assembly, Baltimore, 1988 August.
Papaliolios, C. *et al.* (1988). *George Mason Workshop*, 225.
Particle Data Group (1986). *Phys. Lett.*, **170B**, 1.
Perrier, C. *et al.* (1987). *ESO Workshop*, 187.
Pettini, M. (1988). *Proc. Astron. Soc. Aust.*, **7**, 527.
Pettini, M. Stathakis, R., D'Odorico, S., Molaro, P. and Vladilo, G. (1988). *Ap. J.* **340**, 256.
Pomansky, A. (1987). *Moriond-22*, 739.
Rank, D.M. *et al.* (1988). *Nature*, **331**, 505.
Rees, M. (1987). *Nature*, **328**, 207.
Renzini, A. (1987). *ESO Workshop*, 295.
Rosa, M. (1988). *IAUC*, 4564.
Sagdeev, R.Z. (1988). *Physics Today*, **41**, 30.
Saktara, M. (1988). *Southern Stars*, **33**, in press.
Sandie, W., Nakano, G., Chase, L., Fishman, G., Meegan, C., Wilson, R., Paciesas, W. and Lasche, G. (1988). *IAUC*, 4526.
Sanduleak, N. (1987). *IAUC*, 4318.
Schaefer, B. (1987a). *Astrophys. J.*, **323**, L47.
Schaefer, B. (1987b). *Astrophys. J.*, **323**, L51.
Shapiro, S.L. and Teukolsky, S.A. (1983). *Black Holes, White Dwarfs and Neutron Stars*, J. Wiley.
Shara, M. and McLean, B. (1987). *IAUC*, 4318.
Skinner, G.K. *et al.* (1988). *George Mason Workshop*, 361.
Sonneborn, G. and Kirshner, R. (1987a). *IAUC*, 4333.
Sonneborn, G. and Kirshner, R. (1987b). *IAUC*, 4366.
Sonneborn, G., Altner, B. and Kirshner, R. (1987). *Ap. J.*, **323**, L25.
Spyromilio, J. (1988). AAO Symposium, 1988 September 28.
Stella, L. and Treves, A. (1987). *Astr. Astrophys.*, 185, L5.
Suntzeff, N.B. *et al.* (1988). *Nature*, **334**, 135.
Sunyaev, R. *et al.* (1987). *Nature*, **330**, 227.
Sutton, C. (1988). *New Scientist*, 53 (14 Jan. 88).
Svoboda, R. (1987). *ESO Workshop*, 229.
Teagarden, G. (1988). IAU General Assembly, Baltimore, 1988 August.
Testor, G. and Lortet, M.-C. (1987). *IAUC*, 4352.
Thomas, V., Green, J. and McLendon, B. (1987). *NSSDC News*, **3**(2), 1.
Time (1987). Issue of March 23, 1987.
Tran Thanh Van, J. (ed.) (1988). *Proceedings of the Twenty-Second Rencontre de Moriond*, Vol. 1, *The Standard Model, The Supernova 1981A*, Editions Frontieres. Referred to as *Moriond-22*.

Turner, M. (1988). Private communication.
Turtle, A.J. *et al.* (1987). *Nature*, **327**, 38.
van den Bergh, S. (1987). *ESO Workshop*, 677.
van den Bergh, S., McClure, S. and Evans, R. (1988). *Ap. J.*, **323**, 44. See also *ESO Workshop*, 557.
van der Velde, J.C. (1988). *Moriond-22*, 735.
Vidal-Madjar, A. *et al.* (1987). *Astr. Astrophys.*, **177**, L17.
Walborn, N. (1988). *George Mason Workshop*, 1.
Walborn, N.R., Lasker, B.M., McLean, B. and Lardler, V.G. (1987). *IAUC*, 4321.
Waldrop, M.M. (1987). *Science*, **237**, 25.
Wamsteker, W. *et al.* (1987). *IAUC*, 4410.
Weber, J. (1988). IAU General Assembly, Baltimore, 1988 August.
West, R.M., Lauberts, A., Jørgensen, H.E. and Schuster, H.-E. (1987). *Astr. Astrophys.*, 177, L1.
West, R. (1987). *ESO Workshop*, 5.
White, G. and Malin, D. (1987a). *ESO Workshop*, 11.
White, G. and Malin, D. (1987b). *Nature*, 327, 36.
Woltjer, L. (1987). *ESO Messenger*, 26, March 1987.
Woosley, S.E., Pinto, P.A. Martin, P.G. and Weaver, T.A. (1987). *Astrophys., J.*, **318**, 664.

Index

Plate 1. Top. David Malin's 'before' and 'after' pictures of the supernova, made by combining pictures with the AAT, emphasise the red colour of the nebulae in the region of the supernova (orange halo to the saturated pseudowhite bright image).
Bottom. The 30 Doradus Nebula is the brightest nebula in the LMC and is a region of bright stars in clusters and of dust and nebulae. The supernova occurred in this region and its progenitor was a massive star like all the brighter stars visible in this picture. Photo by David Malin © AAT Board 1987.

Plate 2. Top. Sunrise at the Las Campanas Mountain, Chile. In the centre of the ridge is the University of Toronto observatory, with the 10 in Bruce telescope in the barn-like building.
Bottom. The European Southern Observatory, La Silla, Chile. One of the world's greatest arrays of telescopes is dominated by the 3.6 m telescope building at the rear. © ESO.

Plate 3. Top. Star trails curve in the southern sky above the dome of the AAT in Siding Spring Mountains, Coonabarabran.
Bottom. The Kuiper Airborne Observatory is in an adapted C-141 military transport, here flying over San Francisco.

Plate 4. Top. Ian Shelton, codiscoverer of the supernova, took this 3 hour photograph of it in excellent atmospheric conditions with the University of Toronto's 24 in telescope at Las Campanas on 1987 March 26. © University of Toronto.
Bottom. A 30 minute exposure on Fujichrome by Brian Carter of the SAAO shows the supernova orange on 1987 April 30, southwest of the red 30 Doradus Nebula, here exhibiting its 'tarantula' shape.

Plate 5. Top left. An image of SN 1987A as taken with the Fine Error Sensor (FES) of the IUE satellite. The cross indicates the supernova, which far outshines the other stars in the same region. Analysing this image during an observation of the supernova, the FES holds the IUE spacecraft steady against the drift which is produced by the solar wind.
Top right. The inside of a neutrino observatory is a tank covered with hemispherical photomultiplier tubes, like the facets of an insect's eye. They detect the Cerenkov radiation emitted after neutrinos interact with electrons or protons in the water with which the tank is filled.
Bottom. Diagram from a computer display of a Cerenkov event from one of the neutrinos from SN 1987A detected by the IMB detector. The box structure represents the tank of water and the marks on the walls represent photomultiplier tubes which have been fired by the Cerenkov radiation. The elliptical patch of activated phototubes cuts across the cone of Cerenkov radiation.

Plate 6. Top. Colour-coded plot of optical spectra of SN 1987A as obtained by R.W. Hanuschik and J. Dachs with the University of Bochum telescope at ESO shows how the spectrum changed over the first month (time runs day by day from bottom to top, wavelength from left to right). Very noticeable is the marked fading of the ultraviolet part of the spectrum in the first five days of the outburst (bottom left). In early March more and more absorption features (vertical stripes) began to appear as the supernova cooled. The drift and curvature of the emission at H-alpha (prominent green and blue near-vertical strip to the right of centre) shows how the spectrum progressively came from the slower-moving deeper parts of the supernova as it expanded.
Bottom. Similar colour-coded plot of the ultraviolet spectrum of SN 1987A as obtained by IUE. Courtesy of W. Wamsteker.

Plate 7. Computer simulations by M. Nagasawa, T. Nakamura and S.M. Miyama of Kyoto University show the expansion of the core of the supernova and how in a few seconds the core fragments into lumps which stream radially outwards. These lumps penetrated into the envelope of the supernova and distributed radioactive nickel towards the surface of the star.

Plate 8. Images from the Center for Astrophysics' speckle interferometer used on the 4 m CTIO telescope (courtesy of M. Karovska, CfA). North is at the top, east to the left.
Top left. In 1987 March, the supernova is accompanied by the Mystery Spot. Top right. In 1988 March, the supernova is elliptical, but the Mystery Spot has disappeared. Bottom. Another star observed for comparison is completely symmetrical.